machine design

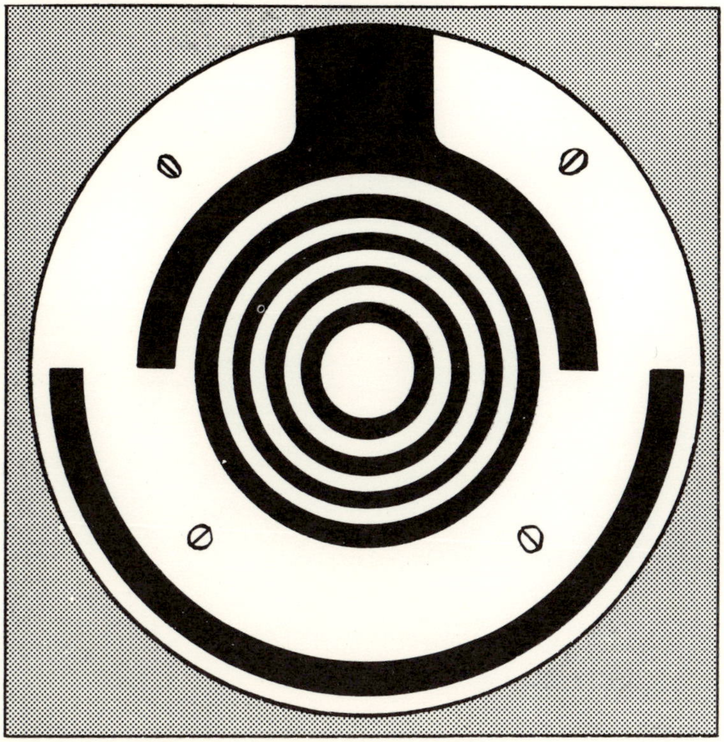

machine design

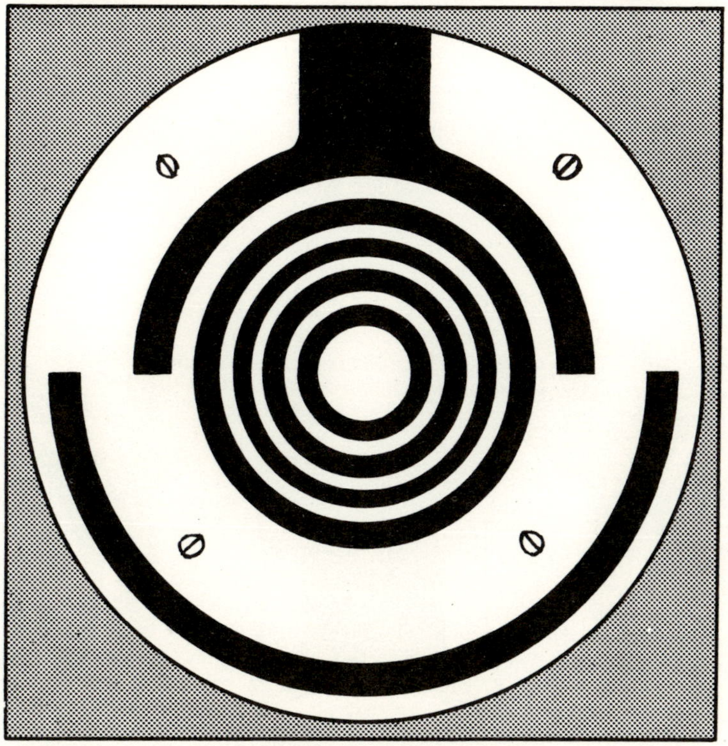

Albert Leyer
Professor of Machine Elements
Technische Hochschule
Stuttgart

English Translation by
D. COOK-RADMORE, BA, FIL
and edited by

SYDNEY URRY
BSc(Eng), CEng, AFRAeS
Professor of Building Technology
Brunel University

Blackie Glasgow London

machine design

Blackie & Son Limited
Bishopbriggs, Glasgow.
5 Fitzhardinge Street,
Portman Square,
London W.1.

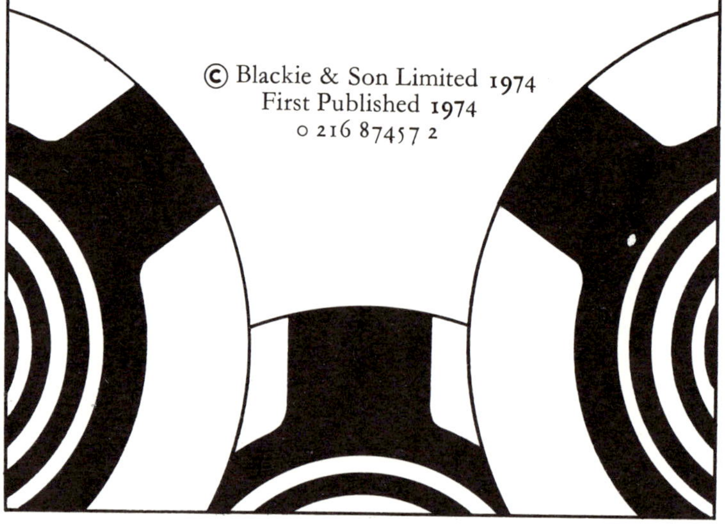

First Published 1974
0 216 87457 2

Printed in Great Britain by
Robert Maclehose and Co. Ltd., Glasgow.

preface

Books on engineering design, like designs themselves, are highly individual. In this one, the author emphasizes the importance of a visual approach to machine design and makes his point by including a large number of illustrations. He also stresses the need for clear objectives in all design work.

Professor Leyer is an experienced designer and an inspiring teacher, and his book is based on his own lecture course in the subject. Throughout, he shows design to be the goal to which mathematics, mechanics and engineering drawing are the means. His book complements the usual range of engineering texts and can be read to advantage by students at any stage of their studies. In addition, he gives clear descriptive accounts of some important topics (such as stress concentration and the torsion of non-circular sections) which are often omitted from textbooks because of their mathematical complexity. In controversial matters—the merits of the patent system, for example—Professor Leyer leaves us in no doubt as to his own views.

In editing this translation I have used SI units for physical quantities and I urge readers to make their own calculations in this system whenever they have the choice. It will be some years, however, before the familiar inch, foot and pound disappear altogether and I have added the corresponding values in these units.

My task has been made easier by the excellence of Mr. Cook-Radmore's translation and the care taken by the Publishers throughout its production. The responsibility for errors remains mine and comments or corrections will be gratefully acknowledged.

Sydney Urry.

contents

Preface

Introduction

PART I. GENERAL ASPECTS OF DESIGN

1 The nature and value of design 1
2 Where does invention end and design begin? 3
3 Design components 7
4 The need for a clear statement of the problem 7
5 The right amount of perseverance 8
6 Design and calculation 9
7 The art of moderation 10
8 The courage to be exceptional 14
9 Design and aesthetics 16
10 Design and skill in drawing 19
11 Standards 22
12 A brief theory of engineering surfaces 26
13 Limitations and changes in the production process 40
14 Control and self-control 53

PART II. GENERAL FORM DESIGN

15 Introduction to general form design 57
16 Elements of design 57
17 The principle of constant wall thickness 61
18 Strong and weak shapes 70
19 The phenomenon of 'lines of force' 82
20 Effect of shape on loading and stresses 83
21 The stress-concentration index (or 'notch sensitivity') of materials 89
22 Designing to match the strength flow 92
23 The shaping of corners and guide facings 102
24 Designing to match fluid flow 108

introduction

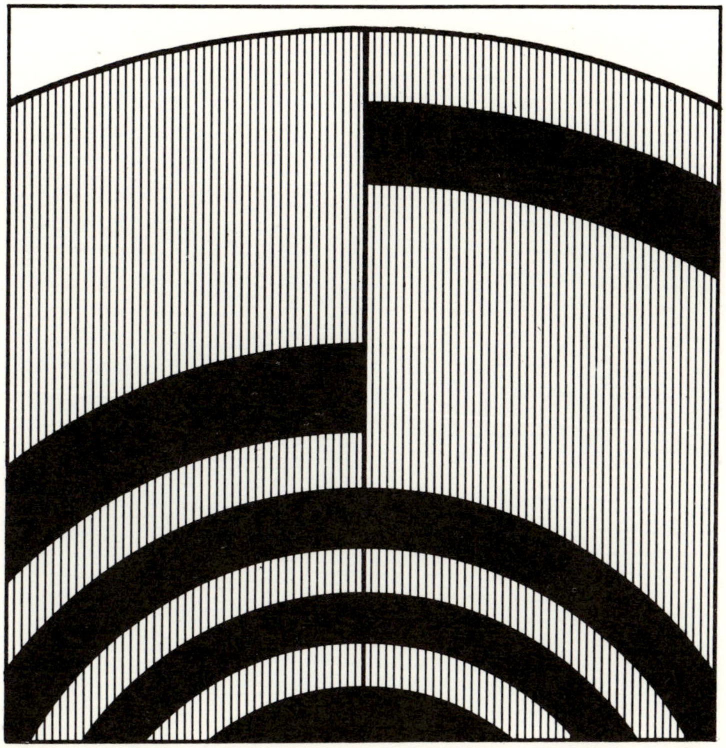

This book has been written at the suggestion of students of the Federal College of Technology at Zurich who wanted a permanent record of the author's lectures on machine design. It will provide students with the general background knowledge they will need later on when they undertake design work in a responsible position.

There are many pitfalls in writing a book like this at a time when design tends to be misunderstood by a large proportion of academic engineers; for this very reason, however, an attempt should be made to give creative work by the engineer the place of importance within his overall activity that is fundamentally due to it. Special thanks are due to the authorities who have given financial support to the preparation of this book.

In dealing with the subject, the author has concentrated on what is universally applicable. The number of illustrations may seem very large; but it is particularly important to appeal to the eye, as well as the mind, and the eye of the designer needs special training. In machine design a great deal is done 'by eye' in the first instance and only later put on a rational basis and committed to paper.

This volume creates a framework on which to build a study of machine design. The various chapters deal with a number of subjects which are often passed over in academic work, but which have a fundamental bearing on the way a student looks at and approaches design as a whole. Other features of the general background of which we have spoken are a knowledge of mechanics, properties and strength of materials and, above all else, a good deal of workshop practice. For our purposes, however, the possession of this basic training is assumed.

part 1
General Aspects of Design

The Nature and Value of Design

There is no doubt that for some time the relationship between the average engineer and design has been a troubled one. His attitude to design is often negative and at best indifferent. While recognizing the necessity for design, he feels little personal involvement in it, and his ideas on what design is about and what significance it holds are essentially vague. When this applies to a responsible decision-making technical specialist, the result is inevitably a disregard and neglect of design compared to the attention devoted to other aspects of the engineer's work. This first chapter is aimed at combating this trend.

Leading members of the profession have described design as the highest and most rewarding activity that technology has to offer. Anyone who knows how to seize the opportunity can give this activity an importance and stature which will be decisive for a whole project or even the whole field of tech-

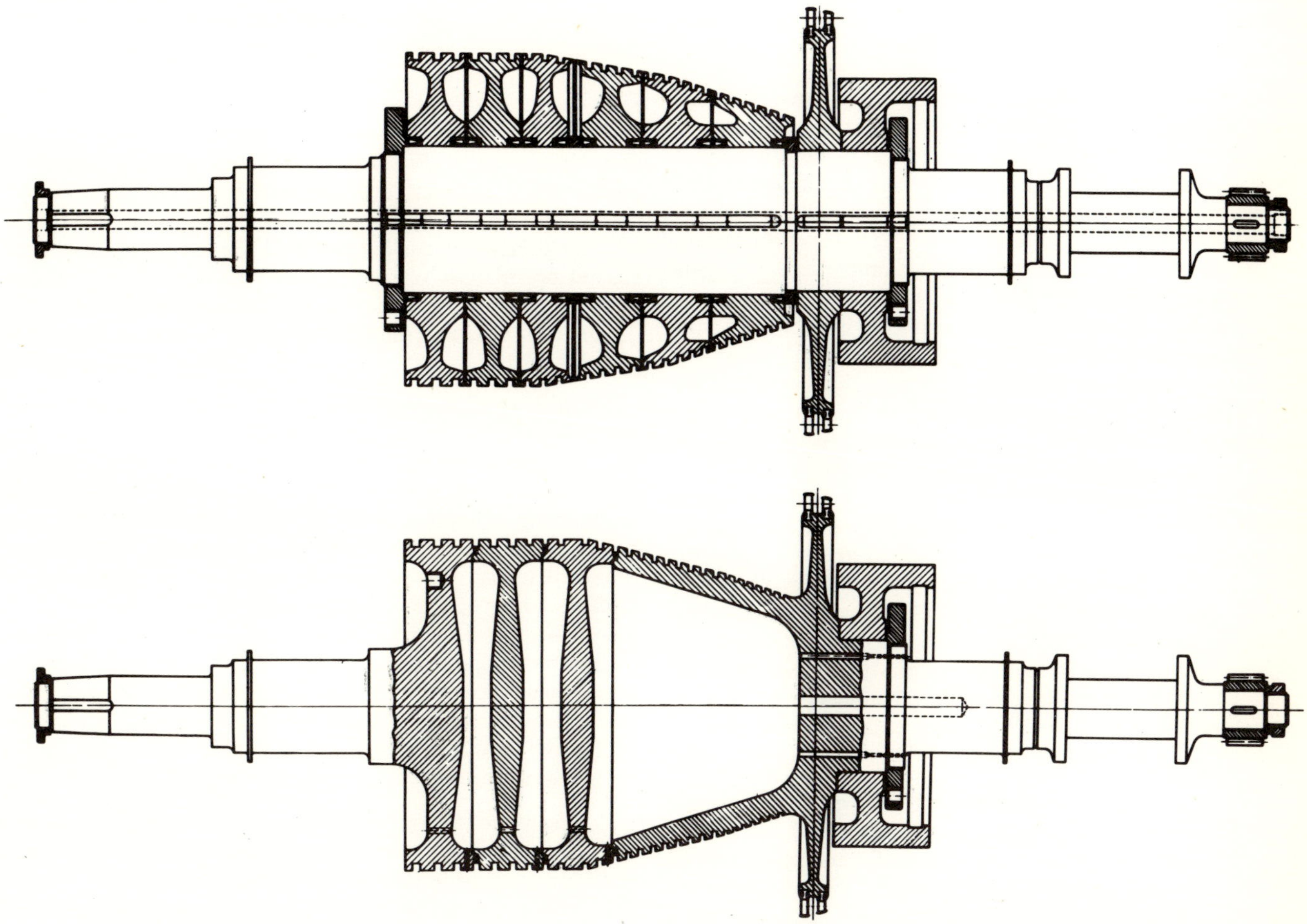

Fig. 1.—For decades the design of steam turbine rotors was based on the same fundamental concept of a continuous shaft with discs attached (top). Rotors were a constant headache to turbine manufacturers, since the gap between the discs weakens the structure considerably. Centrifugal force widens this gap, thus upsetting the true running of the shaft. These difficulties were overcome at one stroke by a new design in which the shaft was discarded, the discs made in solid form and joined by welding around the circumference to form a rotor which is resistant to bending (bottom). (*Drawings: Brown Boveri & Co,. Baden,*)

nology. One meets cases where a firm starts to prosper immediately following the arrival of a talented designer after years of scraping a bare existence. The scope of design depends very much upon the designer. If he is the right man in the right place, no-one can presume to impose petty regulations on him or to set narrow limits to his scope. There is hardly a sector of a firm in productive industry to which the designer cannot make a significant contribution. If he has a proper understanding of what his job means, he will be active everywhere—in the workshop, on the assembly line, on the test rig and even on the customer's premises. I knew one design engineer, a leading man in his time, who worked in shirt sleeves, overalls or a dinner jacket, as circumstances demanded.

It would be hard to overestimate the importance of design to an industrial firm. Whether the firm is successful or not can depend on many factors, but nearly always design plays a decisive role. Consequently when there is an unexpected fluctuation or a fall in business fortunes, the reasons should be sought in the firm's design work as well as in other directions. For some years now it has been commonplace to apply rationalization—scientific reorganization—to the running of a business; and rightly so, for most firms can benefit from such rationalization. The stage has now been reached where the amounts involved are so closely calculated that they can only be expressed in fractions of a second or of a penny. With improvements in design, however, the amounts involved can easily be ten or a hundred times greater. Recently I heard of an engineering works which had to take back a consignment worth £30,000 because of the complete failure of the product, and to offer a 100 per cent replacement which finally cost the firm £50,000. The fault in this case lay in the design and many similar examples could be given. It could take years for 'rationalization' in the firm to make good sums like these—to which must be added the cost of lost prestige. Once there has been a fundamental failure in design, confidence in the firm concerned is lost and a whole network of its best business connections damaged beyond repair.

On the other hand a successful design can lead to unexpected triumphs—and here again many examples could be quoted. Recently there was news of a new type of crane which offered unusually good utilization of space in the workshop in question; this meant that the workshop could be built so much smaller that the saving in building costs was enough to pay for the entire crane installation, all thanks to the inspired work of the designer concerned. A firm with this ability will find it easy to win new orders—Fig. 1 shows a case in point.

It must be recognized therefore that there is hardly any department in an engineering works as important as the design office. It is slowly being realized that the designer is in some ways the key member of the technical team and that only the very best brains should be at the head of the design department. If such people are to be obtained they must be offered appropriate rewards. In the past many businessmen have regarded the design side merely as a necessary evil; there must be a fundamental change in this outlook. In future, design must not rank a long way behind other departments. There has in former times been a tendency to give pride of place to the sales division, on the basis that a strong sales department is far and away the most important factor in a firm's success. But it seems that here, too, a change has taken place; the big customers in the engineering industry are for the most part specialists themselves or are advised by specialists when buying equipment. Such people are now making their presence felt, and pure salesmanship is becoming less and less important in winning contracts. Business negotiations are now conducted mainly on a technical and financial level, so that it is increasingly common for the customer to want to discuss things with the designer himself. Once the technical points have been cleared up, matters of price, conditions of payment and the like are usually quickly settled. The leading part played by the designer is clear, and the sales specialist reverts to his proper place.

An activity as important as design does, of course, make great demands of those who practise it. It is not for nothing that A. Stodola writes, in his book on gas and steam turbines, of the 'sweat and toil' of designing; indeed the persistent unremitting struggle with matter can call for the most strenuous efforts of which the human mind is capable. At the same time it demands the utmost concentration on detail and the understanding of wide-ranging interrelations. Design work forces us to give the most careful consideration to problems, to test, to compare, and to make decisions—always with the realization that we may be making a mistake. Design calls, furthermore, for knowledge which is not confined to a narrow speciality but which ranges far into other fields, such as the technique of mechanical handling (see Fig. 2).

A discipline of this kind teaches clearmindedness and intellectual honesty; it creates an environment in which a good engineer can find fulfilment and develop his talents. Anyone fired with a genuine ambition to meet these high demands and to play his part in Man's creative activity has excellent prospects of advancement in the design world. Despite the attitudes of some people, more and more design engineers are reaching the highest rungs on the ladder of success. There have always been designers, and there are designers today, who are among the best-paid people in industry; they include the 'stars' who can sometimes dictate what they will be paid.

Where does invention end, and design begin?

A technical product does not come about by a continuous process—it evolves through separate and clearly distinguishable stages. The first of these starts with the realization that a certain device or machine is needed—let us say, for example, a machine for converting the available energy of water into work at the rotating output shaft of an engine. First we must be clear in our minds what physical principle is to be employed in the solution of the problem. In the example just mentioned the energy can be transferred by allowing the water either to rotate the shaft by means of suitably-shaped blading (the water-turbine) or to displace the piston of a reciprocating engine and produce the rotation by a crank mechanism.

If the principle involves something new, or something which has not previously been utilized to the end in question, we have *invention*. Every technical product of this kind depends, if we go back to its origins, on invention in some way or other. No matter whether the invention is a major or minor one, important or unimportant, it represents in every case the indispensable prior requirement for the creation of a technical product. If we trace the history of technology from its beginnings, we find inventions at every turn—the first appearance of the wheel, the lever, the wedge and the screw. All of these were inventions, and among the most important inventions of all time. Today they are common property; they are known as machine elements, and every designer makes use of them in building his machines—usually without the thought crossing his mind that he is dealing with the most precious fruits of past inventive genius. That they appeared at a time when the concept of invention did not yet exist does nothing to alter this fact.

It will be appreciated that the choice of principle on which the machine is to work (or, as we are now justified in saying, the invention that is to be used) still provides no guide as to the form the machine is to take. A great deal more is needed for this. If, for example, it is a question of jacking up a motor-car, it is not enough to decide that we shall use the screw principle, or the hydraulic press; putting the principle into effect calls for exact details of form, dimensions, layout and so on. The decision on all these points comprises the second stage in the creation of a technical product, the stage known as *design*. The third and final stage which follows is that of *construction*. Of these three stages, that of design is the most important and decisive. It takes place on a purely intellectual plane, and the design drawings are merely the visible expression of the mental activity.

The view is often held that the truly creative work behind the technical product lies in the invention, but this is not so. It is worth remembering that we get the word *invention* from the Latin roots *in* and *venire*—'to come upon', i.e. 'to find'. Finding is a non-creative act. For example, when it is found (through observation or thought) that mechanical energy can be stored by deforming a piece of flexible metal (the invention of the spring) there is no creativity in the strict sense of the word. The ability to store energy in a piece of metal is not something produced by man; it already existed in Nature when Man, exploring Nature, came across it. Our logical conclusion, despite the generally held view, is that invention is not creative.

If we feel that there must nevertheless be some sort of creative activity in evolving a technical product, then it can

Fig. 2.—The designer is often concerned with transport problems, and it is pointless to go ahead with a design until these have been solved. A load like this giant Swiss-built condenser for a steam power plant in Holland will have to pass through one or two narrow gaps before it arrives at the site. (*Photo: Brown Boveri & Co,. Baden.*)

only be connected with the design stage; and indeed, design is the truly creative part of the process. Something genuinely new occurs only in design, when that which has never been seen, known or thought of by anyone before—i.e. a clear picture, backed up by drawings and perhaps by models as well, of a certain object with a certain shape, size and other characteristics—first comes into existence through intellectual effort. This, and this alone, is creative.

There is a fundamental difference between invention and design; the two cannot be compared and they are subject to different laws. This fact shows itself in many ways. Invention presupposes knowledge—it is the offspring of scientific knowledge and consequently research which if successful always results somehow or other in invention. The discovery of the fact that mercury vapour will only pass electric current in one direction led to the invention of the mercury rectifier. The inventor does not need a great deal of skill; he need not be a technical engineer, therefore, and we can see how it is that many inventions, excellent in some ways, have come from persons with no technical training. Design, on the other hand, can only be carried out by the engineer or technician, since it calls for skill as well as knowledge. Design qualifies as an *art* and the designer is an artist. The timescale of the two processes is also different. Invention always happens instantaneously—all of a sudden the realization is there of how this or that ought to be done. The period during which this realization comes is nearly always brief, just like the discovery of an object. Although inventors often spend days on end over details, this is not really inventing—it is the preparation and gathering of knowledge. There is no correlation between the time taken for an invention to come into being and its significance; whether an invention is a major one can only be judged by its success, not by the way it has come about. With design, things are totally different. The very task of design has a dimension which can be expressed in degrees of difficulty or hours of work. A design project always requires time, usually to be measured in weeks or months. There is never an instance when the design work can be 'dashed off in a moment' and this is why, earlier on, we spoke of the 'sweat and toil' of designing. In this connection we might call to mind a famous saying of Thomas Edison, who commented that his inventions had been 1 per cent inspiration and 99 per cent perspiration. This is seldom understood properly. Edison was not speaking here of invention alone, but of invention plus design; if we give the comment more thought, the only meaning it can have is that the 1 per cent involves the invention process and the 99 per cent the process of design. The invention of the incandescent lamp bulb, for instance, rests on the idea of bringing a conductive filament to white heat by passing an electric current through it and of enveloping it in a protective atmosphere of gas so that it does not burn away. Without any doubt it was not arriving at this idea which gave Edison long days and nights of toil but the working out of how to put it into practice. This was what demanded by far and away the greater part of the intellectual effort—not *what* was to be done, but *how* it was to be done; which is exclusively a matter of design.

The job of inventing is a relatively easy one. This is why there has always been a flood of inventions. To the layman inventions appear as something of inestimable value, but the expert knows that they should not be overrated. If we rummage through the files of patent applications we get the impression that there is hardly anything in this world so plentiful as inventions. Basically they are no more than a kind of raw material from which something has to be made—an invention by itself is nothing. Having the idea of using a rotating steel wire to cut grass (a patent applied for in Germany) does not enable us to do it—for that, an actual cutting machine is needed; and since one has not yet been designed we have no such machine despite the invention. In the same way the knowledge of the principle of an incandescent bulb was not the same thing as the brilliant solution to the problem of illumination provided by Edison—to arrive at that a whole series of problems such as power supply, insulation and sealing had to be worked out. There is every reason to suppose that Edison was not the inventor of the incandescent bulb at all—the idea of such a light-source was too self-evident to many of those dabbling with electricity for it not to have occurred to a large number of people before Edison. What is clear, however, is that Edison was the first to tackle the design of the incandescent bulb seriously and to follow it through to a successful conclusion; this was the 99 per cent of the overall work. If we look at things in this way, Edison forfeits the fame of being the inventor of the electric light bulb, but gains the honour of being its designer, which certainly means a great deal more. Another example of this kind of problem is shown in Figs. 3 and 4.

The inventor and designer are not necessarily two separate persons. It is quite possible to invent something and then work out its design—it is even possible, in doing so, to arrive at further inventions and to incorporate them into the concept of the first. To be able to do this, you need to be an engineer or technician, that is to say, to have scientific training and designing ability. Having the one person as both inventor and designer is the best guarantee of success since it avoids 'handing over'. Experience has shown that a designer will always bring a different approach to his *own* invention than to someone else's—there are, after all, very few people naturally inclined to bringing up other people's children. Consequently this happy combination of inventor and designer in one and the same person should be sought with every means at our

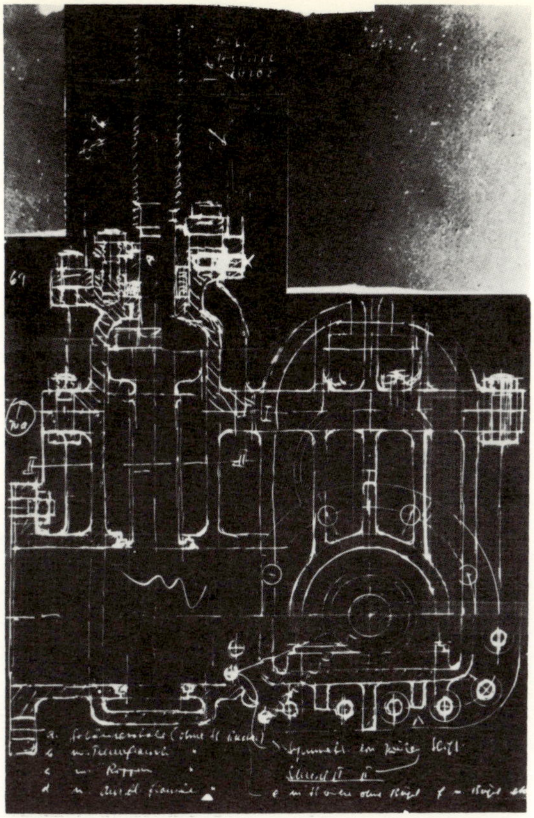

Fig. 25.—Design for a steam valve—an excellent example of freehand technical drawing, with a light and flowing use of line, yet still showing every technical detail clearly.

(Fig. 29), not an obvious adjunct to his many organizational abilities.

Alongside this first kind of drawing, design also uses a second geometric type of drawing, which records what the designer has created in its final definitive form. Its methods are based on descriptive geometry, using orthographic projections in general. Perspective and axonometric drawing are hardly ever used. In contrast to the design sketch, a drawing of this kind is a model of precision and clarity, requiring a full range of instruments—draughting machine, compasses, set-squares, French curves, templates and the like. Ordinarily a pencil will be used, but drawings which will be needed frequently and over long periods are often drawn in ink. The name given to drawings of this kind—*workshop drawings*—indicates their main purpose. Again, unlike the design sketch, they take a long time to prepare; one line is added to another slowly and cautiously, and additions are made with the greatest care. There have been workshop drawings which have taken weeks, or even months, to complete. They have an importance equal to that of the most valuable documents, since they embody what is sometimes years of intense effort by the planning and design departments. Not long ago it was estimated that the amount of information contained in the average A4 (210×297 mm or 8·27×11·69 in.) workshop drawing would take up 80 pages of typescript. The loss of a drawing like this could be a minor disaster for a firm, so those in charge in the workshops, stores and offices using these drawings are only given copies, and never the original. The latter will be kept in a carefully-sited (often fireproof) records store, to which only specially trustworthy persons have access. This documentary nature of workshop drawings is also apparent from the fact that they represent a clear and complete set of instructions for building a piece of machinery from start to finish. Every detail and every explanation of manufacturing method that may be needed is noted, and this can be done in a way which is understood by engineers in any country in the world.

For thousands of craftsmen, the workshop drawing is the

only thing they have to guide them in their work. The drawing is their only means of knowing how and where to cut, or drill, or grind. The slightest misunderstanding—a spline cut in the wrong place or a hole drilled in the wrong place—can make an expensive workpiece useless; those who carry this heavy responsibility therefore deserve to have everything made as easy as possible for them. For this reason, a workshop drawing can never be clear and descriptive enough, and a systematic method should be followed in its production. If a draughtsman is lacking in the necessary breadth of vision and sense of order, a small preliminary draft of the drawing should be prepared, giving careful instructions for the best arrangement of the projections and sections, and the chief dimensions.

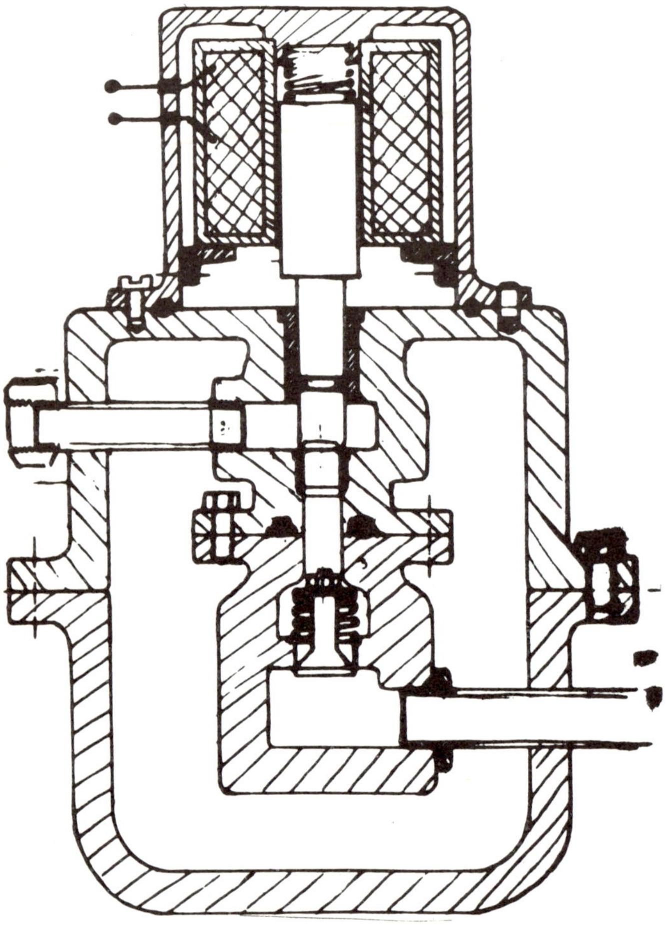

Fig. 26.—Detailed type of technical drawing, with rigid lines, which demands a great deal of care but is of little use for design purposes.

Drawing plays an immensely important role in technology, and the designer must know something about it. He does not need to be a top-grade draughtsman himself, let alone be specially skilled in producing an ink workshop drawing (although one does from time to time come across designers who are remarkably good at this) but he must when necessary be able to step in and tell those actually producing the working drawings the right way to go about things, especially where arrangement, clarity and avoiding ambiguity are concerned.

Standards

One of Man's most distinctive characteristics is that wherever he has settled he has started creating order. Weights and measures and units of quantity and price are introduced, and as he sets about organizing it the whole of life becomes subject to standardization. In the same way standards have come into engineering. As a rule, the young designer has little liking for them. The more independent his personality, and the more pronounced his will, the harder he will collide with standards. He finds that here—long before his arrival on the scene—the whole area has been mapped out, and he is not as free to manoeuvre as he expected. At every turn he comes up against standards, existing arrangements, predetermined dimensions—all of which force him to depart from his original ideas in one direction or another. To him they all represent a sizable obstacle, and he is not at all sure that the disadvantages offered by these standards are balanced out by corresponding advantages; anyone surveying the whole field of standards, and aware of their shortcomings, will have some sympathy for this view.

Standardization is limiting oneself voluntarily to a given selection from the unending variety of possibilities offered by Nature. There is no fundamental difficulty in building a machine using bolts which, from the viewpoint of strength and form, possess ideal characteristics. If, instead of using individually designed bolts like this, we choose bolts with a given thread profile, head and nut shape and a given diameter, we do this not because we are forced to, but rather because we are accepting limitations we have ourselves chosen; and we

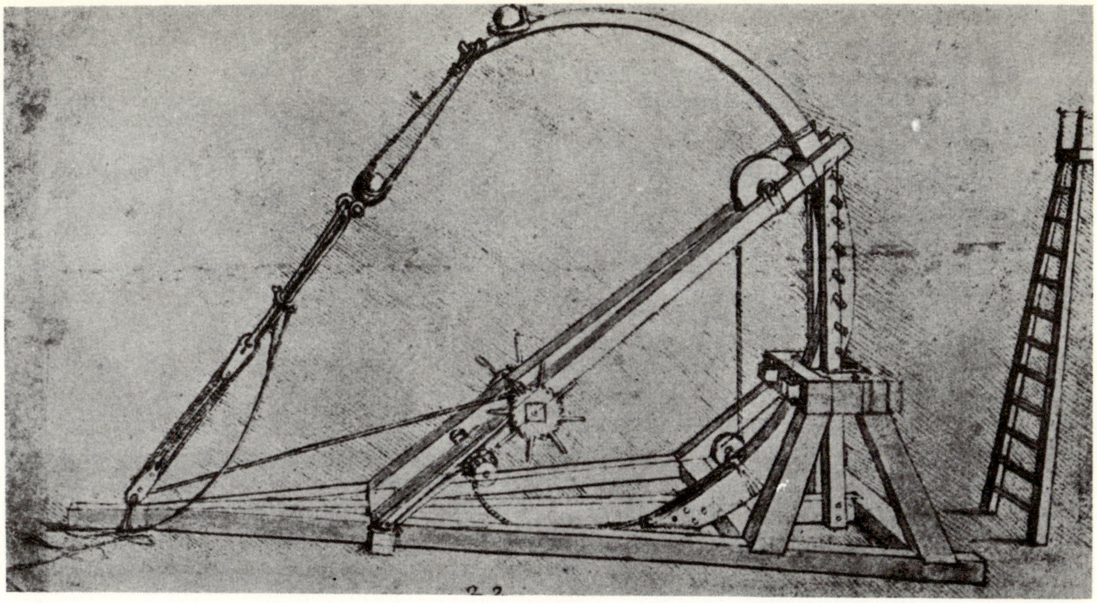

Fig. 27.—Design for a catapult by Leonardo da Vinci. The style of drawing is exemplary, even by modern standards.

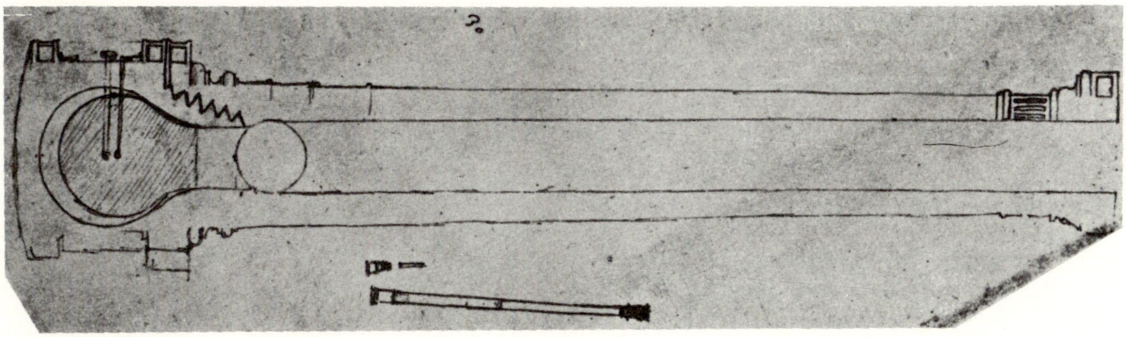

Fig. 28.—Cross-section of a cannon by Leonardo da Vinci. Typical example of design drawing, obviously drawn without straight edge or compasses.

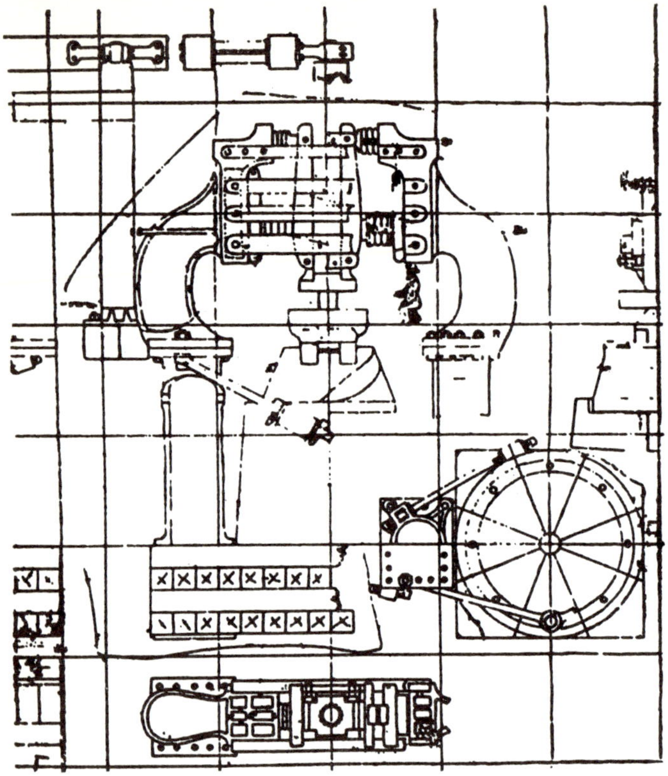

Fig. 29.—Design and layout of the steam-hammer 'Fritz' drawn freehand by Alfred Krupp in 1859.

have chosen them not because they offer advantages in every respect but because their advantages outweigh their disadvantages.

Arriving at a standard such that the advantages outweigh the disadvantages to the greatest possible extent is an extremely exacting task, which has an immensely important bearing on machine design and calls for a great deal of knowledge and skill. This means that only competent specialists should ever be entrusted with the drawing up of standards; what they consider to be right, and decide upon, can be binding for many years on innumerable industrial concerns. For an industrial country a standard which turns out to be a mistake can have effects amounting almost to an economic disaster.

Probably the most difficult aspect of arriving at a standard is that arrangements have to be made for the future without knowing what course future developments are going to take. It has happened more than once that decisions on standards have had to be retracted and modified some time later, and it can be appreciated that this can sometimes involve substantial losses when stocks of parts, tools or even whole machine-tools become unusable overnight.

Because of the great importance of standards, all industrial nations have set up extensive standards organizations, such as the British Standards Institution and the American Bureau of Standards, which often have dozens of sub-committees. These national standards offices belong to a vast world-wide organization, the ISO (International Standards Organization), which has its headquarters in Geneva. Its activities can be counted among the greatest pioneering achievements of technological history. Each national standards organization publishes 'specifications' which today number many hundreds. The main subjects with which they deal (as far as machine design is concerned) cover technical drawing, materials, semi-finished products, bolts and screws, keys, fixing bolts, fittings and the like. Taken together they present an enormous range which is very difficult to grasp quickly, and because of this a study of standards as a subject in itself is essential for the designer. He will not, of course, need to know the details, but he ought to know of all the items that have been made the subject of a standard, at least within his own field. Since this is a requirement that is not met by a great many designers, many firms have—in addition to their standards offices—a standards supervision department, which checks that they comply with the standards. These departments are an extremely useful institution, even though they are not popular with the majority of designers.

The most important mathematical aid to standardization is provided by *preferred numbers*. The limitation to given shapes and sizes we have mentioned finds its practical expression in a limitation to given numbers for such things as dimensions, relationships, speeds of rotation and the like. Our normal system of numbers (1, 2, 3, 4, etc.) is based on the constant addition of one unit at a time, and forms an arithmetical progression. With the introduction of the decimal system, this arithmetical progression had superimposed on it a geometrical progression with a step of 1 : 10. This duality in our numerical system presents us with a substantial obstacle when it comes to standardization. In choosing numerical values, we try to take the most obvious numbers from each system—for instance the numbers 2, 4, 6, 8 from the arithmetical series and the steps 1, 5, 10, 50, from the geometrical series; as a result the series of numbers we end up with is heterogeneous—that is to say, it has a partly arithmetical and partly geometrical structure. It very soon became obvious that the arithmetical series was unsuited to standardization work. If, for example, the lengths of splines are to be standardized over the range 20–200 mm, then the spacing 20, 25, 30, 35 mm at first seems ideally suited. When we get to the greater lengths above 100 mm, however, this spacing is found to be too narrow. Intervals of 100, 105, 110, 115 mm and so on offer a far greater range of possibilities than is needed in practice, and, for these greater lengths, intervals of 100, 125, 150, 175 mm would be found perfectly adequate. This brings us up against a phenomenon we might call the law of constant proportion: we feel that the interval, Δs, should bear a constant relation to the value s, itself, i.e. $\Delta s/s$ should be constant. It will easily be seen that this is the case when the quotient of two consecutive values is also constant, i.e. when the figures in question make up a geometrical progression. Fig. 30 compares two series of pedestal bearings, one spaced at arithmetical intervals of linear size and the other at geometrical intervals.

The question now is which series should we take. It will be appreciated that for practical engineering purposes we are only interested in a series that will contain at least all the tenth powers of our decimal system; besides this it ought to have the half and quarter values of each power of ten and, if possible, the eighth value as well, i.e. the numbers 5, 2·5 and 1·25. Furthermore, it would be desirable for the series to have the straight multiples of the unit, i.e. the numbers 2, 4 and 8. We would then have, from 1 to 10:

1   1·25   2   2·5   4   5   8   10

This provides a provisional list of what we would like, and any practising engineer will see how sound it is. It can now be seen that there are some wide intervals which are about twice as large as those between the pairs of numbers with small intervals. If we divide each of these larger intervals into two equal parts by inserting the appropriate numbers, we get the following series:

1   1·25   1·58   2   2·5   3·16   4   5   6·32   8   10

This is not exactly a geometrical progression, since the ratio of consecutive numbers varies between 1·25 and 1·266. In a single decimal range we have, by coincidence, ten steps or intervals: an exact geometrical series with corresponding numbers would have an interval of $\sqrt[10]{10}$, that is to say 1·2589. The series would therefore run 1 . . . 1·2589 . . . 1·5849 . . . 1·9953 . . . 2·5119 . . . 3·1623 . . . 3·9811 . . . 5·0119 . . . 6·3096 . . . 7·9433 . . . 10. No-one could say that it would be an easy task to work out any sort of standards using this series of numbers. Numbers which are all infinite decimals are unsuitable, but we can see that, with three exceptions, they lie very close to round numbers, and precisely those numbers that we wrote down in our provisional 'list of wants'. The fact that they are not in absolutely regular proportions is virtually immaterial,

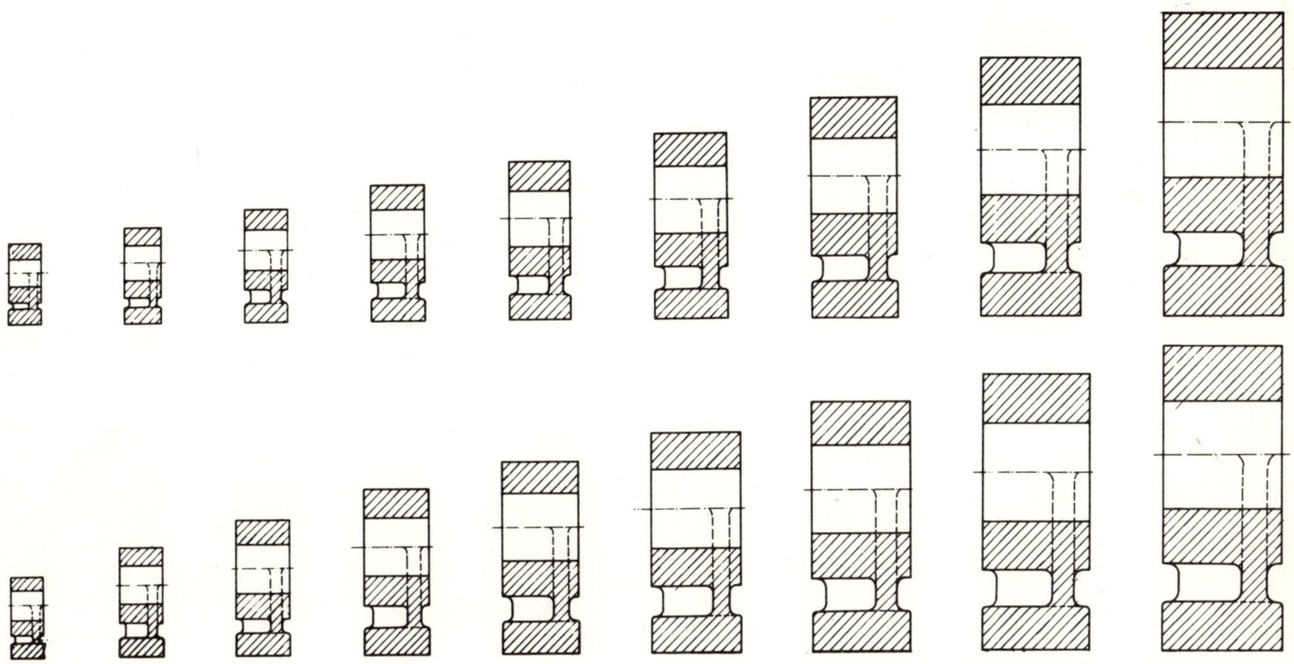

Fig. 30.—Two series of pedestal bearings, one arranged in sizes following an arithmetical progression, the other in geometrical progression.

and where the three exceptions are concerned, these can be rounded off to 1·6, 3·15 and 6·3. This gives us, for practical use, the series:

1 1·25 1·6 2 2·5 3·15 4 5 6·3 8 10

The series can, of course, be extended indefinitely in either direction. Of all the geometrical series that can be fitted into our decimal system this one offers the most advantages, and represents a gift of Nature for use in standards work.

It is interesting to note that it has been discovered independently on a number of occasions. Evidently the reasoning given above, which leads to this series, is a very obvious train of thought. The series was soon put to use for drawing up all kinds of standards; one of the first instances was when Renard, a French army colonel, compiled standards for the guy-ropes of captive balloons. The series was called the Renard Numbers in his honour and indicated by the symbol R_{10}, the figure 10 showing the index of the root.

Years ago the ISO decided to introduce this series officially as the standard series, and recommends it to the national standards organizations for the widest possible application in all types of standardization (see B.S. 2045:1965). Its introduction has, however, met with a fair degree of resistance, the reason for which is not easy to understand. It must be assumed that the advantages of preferred numbers are still not well enough known; the advantages become more obvious as wider use is made of them. Some firms recognized what was to be gained from the use of preferred numbers at an early stage, and went over to a consistent application of these in their own internal standards on a grand scale, for such products as pump impellers, coil springs and control gear. These firms claim that the advantages that have accrued are very considerable.

If the dimensions of standard parts are consistently aligned to preferred numbers, then the preferred numbers can also be transferred to calculation work, since all products, quotients and whole-number powers of these will themselves be preferred numbers. Since π is nearly equal to 3·15, this, too, can be regarded as a preferred number, and as it enters into all calculations involving circles (such as the determination of circumferences, circular areas, moments of area and moments of inertia) we are once again operating within the range of preferred numbers.

The logarithms of the R_{10} series for the arithmetical progression are 0, 0·1, 0·2, 0·3 and so on; this fact can be used for finding the preferred number with the help of a normal slide-rule, on which the logarithms unit has ten divisions. They can also be obtained from the specifications on preferred numbers published by the national standardizing organizations (see, for example, British Standard 2045: 1965—*Preferred Numbers*).

Where the intervals of the R_{10} series are not suitable, new series can be formed by leaving out every second number or by further subdivision. This gives, for example, the series:

R_5	1	1·6	2·5	4	6·3	10	16
R_{20}	1	1·12	1·25	1·4	1·6	1·8	2
R_{40}	1	1·06	1·12	1·18	1·25	1·32	1·4

(continued)					
	25	40	63	100	etc.
	2·24	2·5	2·8	3·15	etc.
	1·5	1·6	1·7		etc.

From these series it is possible to develop others by removing every *n*th term, thus meeting any conceivable requirement. Because of the advantages they offer, preferred numbers ought to have been used in mechanical engineering from the start, to serve as a general basis for all standardization. A standard for bolts based on preferred numbers, for instance, would have offered immense advantages, and the same applies to a standard for ball-bearings. Since, however, the standards for these were laid down at a time when the system of preferred numbers was still very little used, they have no connection at all with it. We can only hope that there will be a basic change made when these standards are due for revision.

A brief theory of engineering surfaces

The engineer is interested in two aspects of a surface—its position and its form, and both of these have to be considered in putting his design into effect. The position of the surface will be noted on the workshop drawing by means of dimensions, and the form by conventional symbols or brief notes.

Fig. 31.—This gives some idea of the difficulties faced in making an accurate measurement of a metal surface (fracture surface of a steel connecting-rod, approximately full size).

a

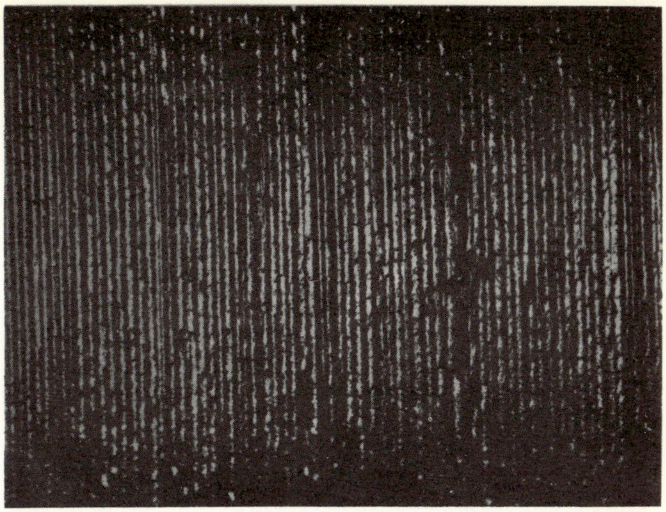

b

Fig. 32.—Photograph (*a*) shows a raw cast-iron surface, magnified about 20 times, contrasting with the split structure of the machined surface in (*b*).

The functional effectiveness of a machine part will depend to a very large degree on whether the dimensional limits and surface finish match the requirements, and measuring the position and evaluating the form of a surface in the modern technical understanding of the term is one of the most difficult problems that has ever faced the engineer (Fig. 31).

Where specifying the position of a surface is concerned, it would seem as if the introduction of an international system of tolerances and the advent of a vast number of up-to-date measuring instruments and measuring processes had overcome the main difficulties. But when the question is studied more closely, it can be seen that this problem is bound up with that of the form of the surface—that the question of exactly where a surface is can only be given a meaningful answer if we take the profile of that surface into account; and the difficulties that crowd in when we think about surface form are so numerous that they seem to defy solution.

Surface Form

In the early days of mechanical engineering the form of surfaces was given hardly any thought, and the designer accepted them just as they came from the normal manufacturing processes, such as turning, stamping, casting and so on. Only in exceptional cases did he give any special instructions on how a surface was to be finished. However, sad and bitter experience began to show that the nature of the surface can have a decisive effect on the behaviour of a machine part; rough surfaces in a bearing, for example, can sometimes prove more resistant to wear than smooth surfaces, while on the other hand rough surfaces have a marked detrimental effect on the fatigue strength of bolts. Tiny surface irregularities in the bottom of a groove may result in fractures at apparently negligible loads. It has been found that such strength effects can vary as much as 50 times, depending on the surface finish. A striking discovery is that quite small irregularities can often have a result out of all proportion to their size—one turbine manufacturer had constant trouble with bearings over a period of years which stopped immediately when small—indeed, hardly detectable—highspots were removed from the bearing surface with an oilstone. A further example is provided by the extreme brittleness of graticulated glass. Cases like this force the engineer to give a good deal of attention to the phenomenon of surface finish.

A study of the problems of workpiece surfaces shows that these are not only of a geometric nature—they involve the materials themselves as well, since it is not just a matter of considering the form of the top surface; the underlying stratum is also of major importance. Both the *structure* and the *texture* of a workpiece at the surface of the material will depend to a very large extent on the way this surface has been produced. Rolling and stamping processes alter the crystalline structure and the relation of the crystals to each other, and result in a schistous structure which gives excellent strength characteristics. With turning, grinding and milling, on the other hand, the crystals are destroyed and torn apart and a ripped-up layer results even when the cutting tools are in first-class condition (Fig. 32). A vast number of tiny fissures penetrate the workpiece, and have a very adverse effect on its strength.

For this reason a thread which has been produced with a cutting tool is weaker than one which has been rolled. It is common knowledge that casting produces a highly-resistant surface. The rapid cooling at the surface gives the carbon no time in which to crystallize out and the result is the solid iron-carbon compound cementite, which is also the main component of the wear-resistant top layer of hardened steels. Many heat treatments also bring about changes at the surfaces of the material, such as cementation through case-hardening, quenching and tempering. Other well-known processes are surface alloying with copper and silver, and coating with nickel and chrome; these not only fill in the fissures that have come about during the machining process, but also alter the

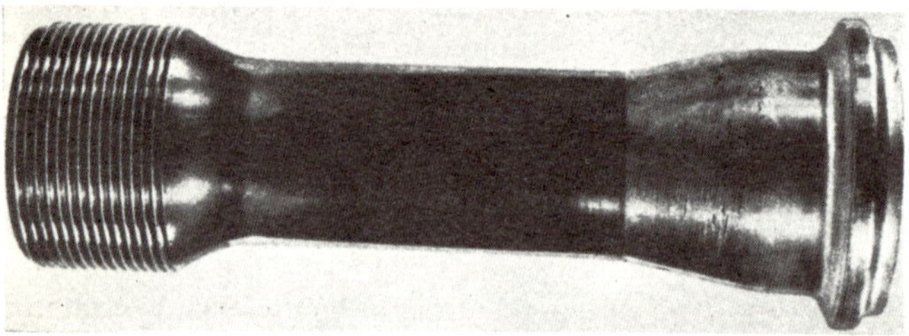

Fig. 33.—Bolt for an aero-engine crankshaft, with Hirth serrations. The head is silver-plated, the centre section copper-plated, and the thread end lead-coated.

crystalline structure through partly involving these metals as components in the alloy. An interesting example of how the properties of a machine part can be improved by depositing thin layers of metal is provided by the bolt seen in Fig. 33. But the principal difficulty in this problem of surfaces lies undoubtedly in the topographical form. Fig. 31 will give an idea of the snags that face us when we try to measure a surface and give an accurate description and picture of it, while Fig. 34 shows the wonders and secrets that are sometimes uncovered during research on surfaces. The physical behaviour that might be expected from such a surface with regard to such things as heat radiation, absorbent capacity or friction would fill a book.

The design engineer's first need is to break down the form of a surface into its various elements and to arrive at a clear and characteristic picture of these. If he wants to have an idea of the quality of the surfaces of a component, it will not be enough to scratch it with a fingernail and go on the subjective impression gained. On the other hand it is easy to see that we shall never be able to consider every tiny detail of the surface; what we have to do is to limit our analysis and study of a surface to what is needed for a satisfactory understanding of its physical behaviour. So far, however, we have not made much progress in this direction.

Until quite recent times the means of indicating surface finish was merely the three symbols ▽ ▽▽ ▽▽▽, the first of which means a coarse finish, the second a smooth finish, and the third an ultra-smooth finish. It will be obvious that this was not enough to cover the needs of mechanical engineering, not only because it took into account only some of the machining processes (and made no allowance for grinding, polishing, lapping, sandblasting, etc.) but because these indications themselves allowed of the widest variations. So it is not surprising that new systems for indicating surface finish had to be evolved, the Swiss VSM system probably being the most detailed of these. Every possible way of treating and forming a surface has been included in this system, tabulated and numbered. The following list shows how the 99 numbers specified in the VSM system are grouped:

01–09 = coarse finishing
10–19 = machining finish
20–29 = heat finishing
30–39 = embellishing finish
40–59 = coating
60–69 = bonding
72–74 = sealing
80–99 = various

In many countries indications have been limited to brief notes, such as honing, copper-plating and so on, but this does nothing

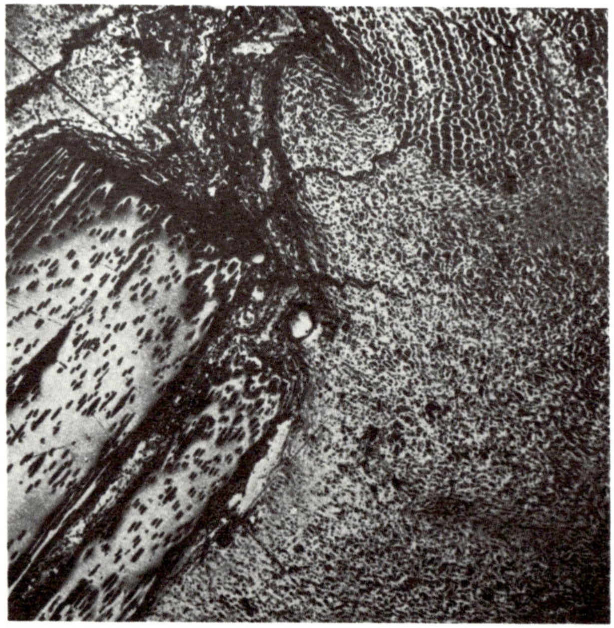

Fig. 34.—Surface of coal, magnified 50 times (actual area about 4 mm²) and showing a wide variety of features, some of which betray its vegetable origin; the cellular structure (top right) is still obvious in spite of the age of the sample. The lump of wood coming in from the left also shows clearly the remains of wood cells (C. Zeiss, Oberkochen).

to make clear the fact that there are basically two problems involved—that of describing the geometrical form on the one hand, and the physical-cum-metallurgical character on the other. That this system fails to cope with the first of these problems (because a wide variety of surface characteristics are lumped together under a single symbol) is evident in that a number of firms have adopted the use of geometric surface roughness indicators, which go much further in identifying the different kinds and degrees of finish. These indicators are graduated in degrees of surface roughness, and it seems that the practical results obtained are very satisfactory.

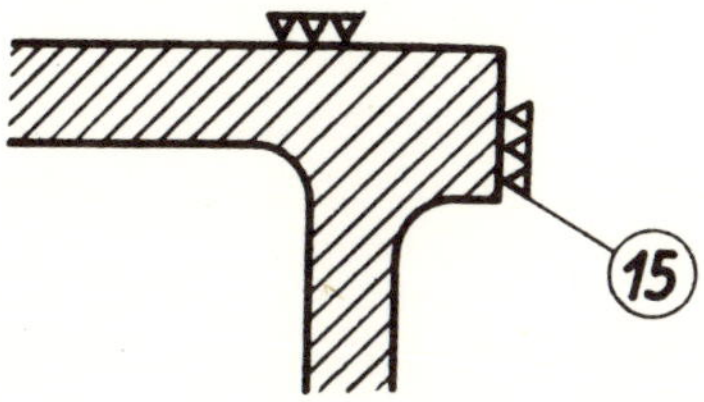

The idea of making these surface roughness indicators into an official standard and including this among the existing national standards is an obvious one, but it has been realized that any moves in this direction must be preceded by the creation of a basic standard on what might be called the 'topography of engineering surface finishes'. This was tackled some years ago by the standards organizations in various countries. The extent of the undertaking can be judged from the fact that, in the Swiss specification, for example, it entailed the definition of 50 new terms which had not featured in the engineer's vocabulary before. In the United Kingdom the work was undertaken by the British Standards Institution and published in B.S.1134: 1961 *Centre-line-average height method for the assessment of Surface Texture.*

This work of standardization was based on the fundamental realization that it is impossible ever to arrive at the true form of a surface, and that we shall always have to be satisfied with a greater or lesser degree of approximation in specifying it. The extent of the fidelity achieved depends on the method used for recording the form of the surface, several alternatives being available to us. The first of these involves the 'scanning' of the surface with the tip of a stylus, the deflections of which are transferred through a lateral movement of the scanning mechanism and suitably magnified. The resulting traces show us that there are two main features making up the form of a surface: one which extends over fairly large areas of the surface (e.g. flattish bumps, depressions and waves), and another covering smaller areas, superimposed on the first, which can be regarded more as 'roughness'. It is, indeed, difficult to make definitions which distinguish clearly between these two forms. As, however, there are very many types of surface met within engineering practice which exhibit this dual nature, it was decided to make this basic division and to refer to the first as *waviness* (or secondary texture) and the second as *roughness* (or primary texture). See B.S.1134 for definitions of these terms.

The two components can be recorded by means of two styli or feelers, the first of which is shaped like a skid with a fairly large radius of curvature and passes over the finer irregularities to record mainly the larger extensive shaping defects, such as waviness, while the second stylus has a fine tip; this is attached to the first, so that it moves with it to record only the roughness component. This feeler system is mounted on an arm, and the

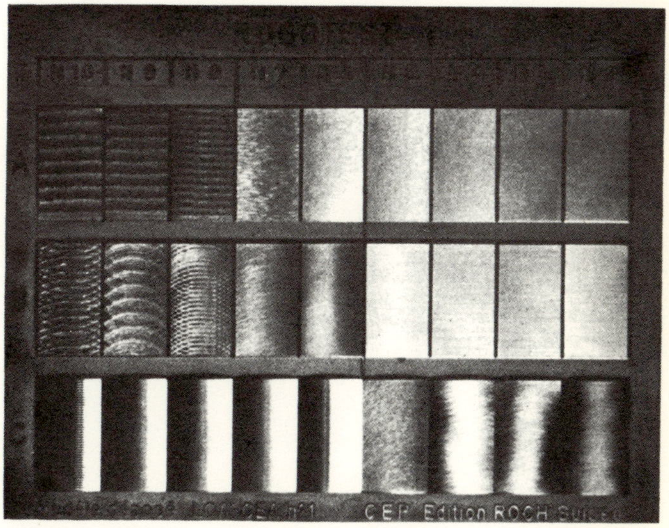

Fig. 35.—Surface finish indicators for machining metal with cutting tools, giving samples of hobbed, end-milled and turned surfaces arranged in a geometrical series of surface roughness.

deflections are passed along this arm from the test surface to be recorded on a strip of chart paper. The curve from the first stylus is traced as an envelope profile or 'evolute'. In B.S.1134 it is termed the 'error of form' profile. Fig. 36 shows the two traces, with this evolute at the top and the roughness profile beneath it; below these again, we see the two combined, forming the actual profile. The way in which the roughness is depicted here does not give a true idea of the roughness defect component, which appears only as valleys in the surface. In fact the roughness rather consists of both plus and minus variations from a smooth surface, and these variations we can get from plotting the roughness defects against a centre line, whose position is obtained from a calculation of the mean value R_q using the formula:

$$R_q = \frac{\int_0^l y\,dx}{l}$$

where x and l are measured along the surface. This reference to the centre line gives a far more accurate picture of the actual roughness (Fig. 36 *a–d*). An average figure for this roughness value will be enough for making an initial assessment of a surface, and even for many of the calculations needed. This average figure R_a is obtained by means of the formula:

$$R_a = \frac{\int_0^l (y - R_q)\,dx}{l/2}$$

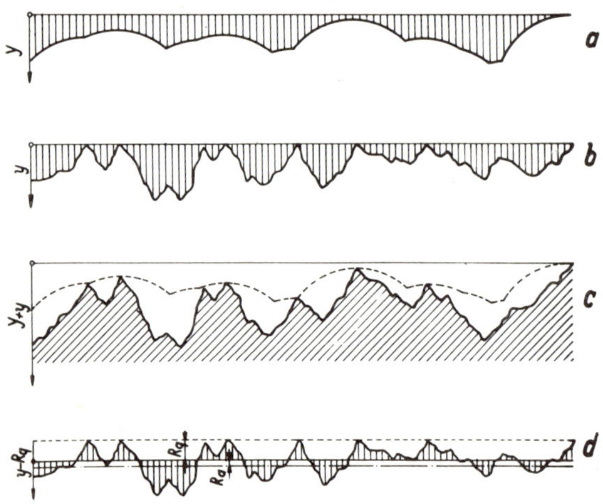

Fig. 36.—Representation of section profiles.
(*a*) envelope profile or 'evolute', *Y* form defect,
(*b*) *y* roughness defect,
(*c*) *Y*+*y* total defect, giving the actual profile,
(*d*) roughness defect superimposed on average roughness depth R_q. Average roughness value R_a.

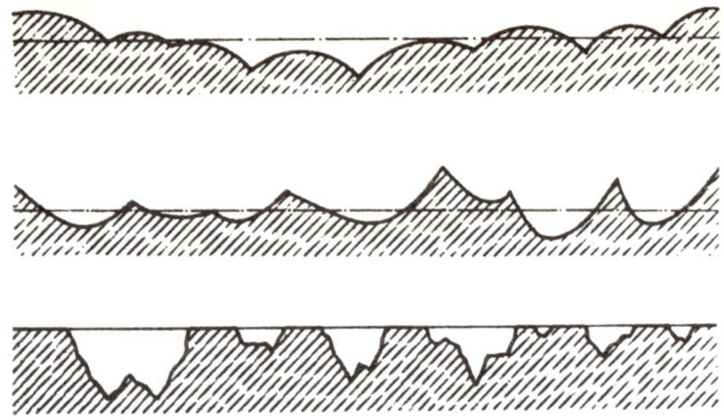

Fig. 37.—Various types of surface, showing the need for additional ways of specifying the surface.

where the integral applies only to those parts of the curve lying above or below the centre line that have an equivalent surface area. This value R_a is adopted as a main concept in the surface roughness standards of the VSM, where it is termed the 'roughness value'. In the British system (see B.S.1134) a *centre-line-average* height is defined. Many experts who are concerned for a large part of their time with the assessment of surface finish maintain that it is possible to detect even slight differences in the average roughness value by merely scratching the surface with a fingernail. It is, of course, not possible to give a comprehensive picture of the roughness characteristics of a surface by means of the R_a value, and it is quite easy for surfaces to have the same average roughness figure and yet have very different technical and physical features. Fig. 37 illustrates three kinds of roughness and will make this clear. It would probably be difficult, merely by introducing further characteristics of a geometric nature,

Fig. 38.—Surface testing instrument by Taylor & Hobson (Leicester).

ROUGHNESS CLASSIFICATION ACCORDING TO VSM

Roughness classification	N12	N11	N10	N9	N8	N7	N6	N5	N4	N3	N2	N1
Average roughness value	50	25	12·5	6·3	3·2	1·6	0·8	0·4	0·2	0·1	0·05	0·025
Symbol	▽			▽▽			▽▽▽			▽▽▽▽		
Type	Course			Average			Fine			Finest		

to cover the typical features of a surface well enough for the operator in the workshop to be able to work from the result. In spite of the considerable efforts that have been made, the basic problem still seems to be just as far from being solved—the problem of finding a simple but clear way for the designer to formulate his instructions to the workshop and a way of checking whether and how these instructions have been carried out. The surface roughness indicators do, however, seem to have at least partially filled the gap.

In the meantime the Swiss VSM have gone on to expand further the standard which deals with the indication of surface finish by means of three machining symbols. This is based on the newly-coined concept of the average roughness value R_a. Twelve classes of surface roughness have been introduced, with the degree of roughness arranged on the system of preferred numbers. These roughness classes are split into groups of three, to which are allotted the triangular symbols already described plus a fourth consisting of four triangles.

To show which of the three roughness classes is intended by the group symbol in any particular case, the number of the roughness class in question is added, with the exception of the coarsest in the group. For instance we write N11, N7, N5 and N1, but in the case of classes N12, N9, N6 and N3 only the machining symbol for the group is shown. Taken in conjunction with the roughness surface indicators mentioned earlier, this does provide a basis for a practical approach to the problem of surface finish. An important factor in this is that the most commonly-used roughness indicators already make use of the roughness classes shown in the table given above, and adopt the same numbers and symbols (Fig. 35). See also B.S.2634 *Roughness comparison specimens.*

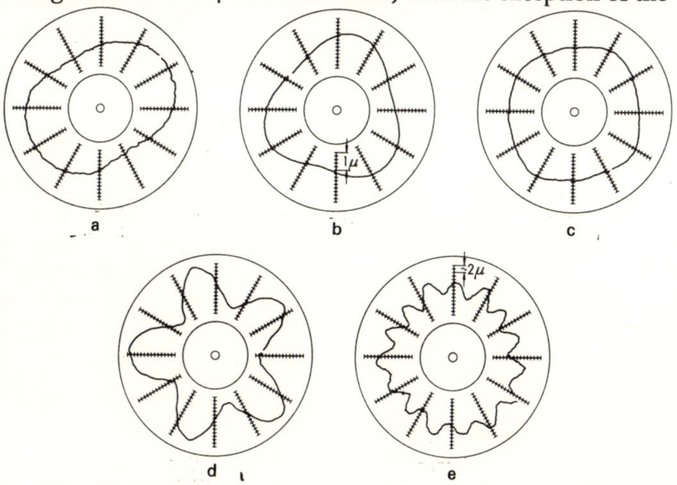

Fig. 39.—Profiles recorded from bearing parts with the Talyround: (*a*) inner ring of a parallel roller bearing, (*b*) ball, (*c*) inner ring of a taper roller bearing, (*d*) taper roller, (*e*) outer ring of a taper roller bearing. All these items were rejected at inspection (SKF).

As the problem of surface finish has come more and more to claim the attention of the engineer, most firms have set up departments to go into the problem. Indeed it is difficult to imagine how long-term success in the fields of ball-bearings, optics, precision gauge manufacture and gear manufacture can be possible without an intensive study of the problem of surface finish. There has been a corresponding increase in the importance of equipment for testing this, and nowadays there is no large firm that does not have at least one such test apparatus in operation.

There are various ways of recording the shape of a surface with these test machines. We have already mentioned the mechanical type, which includes the equipment made by Taylor and Hobson (of Leicester, England). One of the most common models of this is shown in Fig. 38, and Fig. 39 reproduces charts made from tests on bearings with this machine. The roughnesses of the surface shown here have already largely been removed by grinding and polishing, leaving only the waviness. It is precisely these, however, that have a very decisive effect on the way a ball-bearing runs; they are hard to overcome, since they result from inaccuracies in the manufacturing machines. Further examples from a Talysurf recorder are shown in Fig. 40. Profiles *a* and *b* betray clear periodicity, which is undoubtedly connected with the tool feed. The turned surfaces of trace *c* has a less marked period, while the ground surface of trace *d* follows a completely irregular pattern. A further type of surface recorder is seen in Fig. 41; this is the Perth-O-Meter (by Hommelwerke of Mannheim) which allows the roughness defect and form defect to be separated as the recording is made. Fig. 42 illustrates a recording made with this equipment.

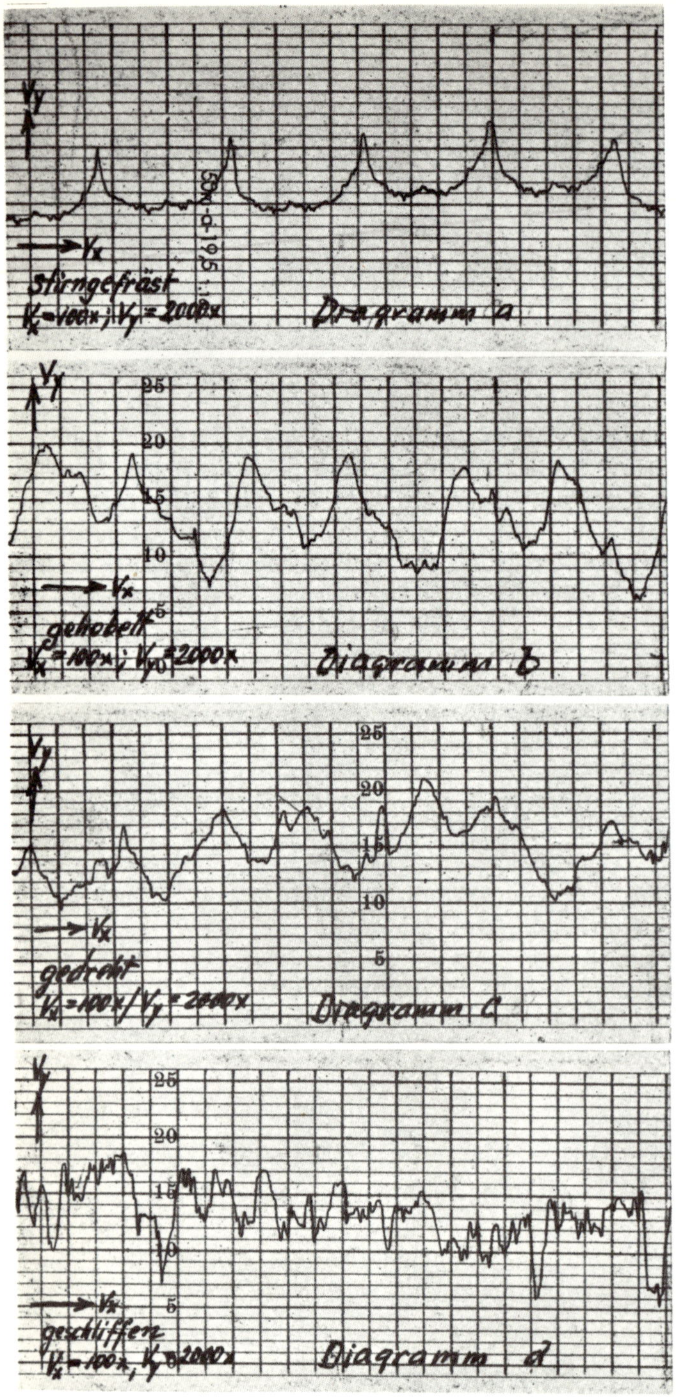

Fig. 40.—Section profiles recorded from various surfaces with the Talysurf.

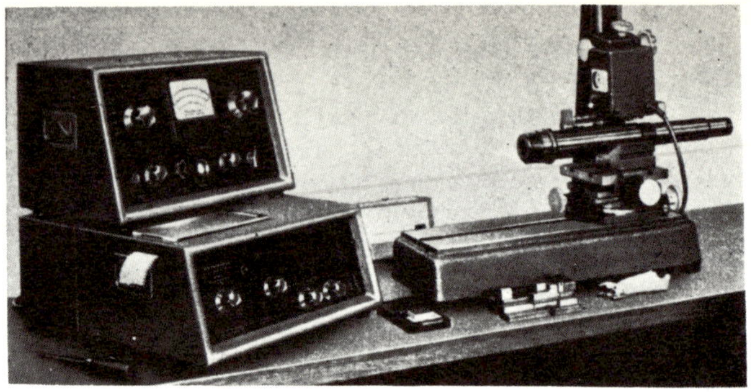

Fig. 41.—Perth-O-Meter. Recording instrument for surface profiles which automatically separates the form and roughness defects. Developed and built by Dr. P. Perthen and Messrs. Hommelwerk, Mannheim.

The first of the optical methods that should be mentioned is the simple process using oblique lighting; while this admittedly only gives a view of the surface rather like an enlarged aerial reconnaissance photograph, it can provide extremely useful information (Figs. 31, 32, 43 and 44). A direct sight of the surface like that provided in these pictures is, because of its vividness and emphasis, superior to that of any indirect method of recording.

Probably the oldest optical method for inspecting surfaces is based on interference, and was developed by Carl Zeiss (Fig. 45). The surface to be inspected is covered with a glass plate, and a beam of monochromatic light is directed vertically onto it. Part of the light is reflected by the lower surface of the glass plate, and the remainder passes across the gap below the plate onto the test surface, when it is reflected back. This second beam of reflected light follows the first with a slight time-lag, and—depending on the size of the gap—either increases or diminishes it; the result is a dark-and-light effect running across the test surface in the form of lines. The nature of the surface can be judged from the closeness and pattern of these lines, just as the topography of an area of countryside can be judged from contour lines on a map. In Figs. 46 and 47 we see the same surface recorded first by oblique lighting and secondly by the interference method. It can be seen from this how the latter method will reproduce a very complex surface picture with precision, while the former picture is quite difficult to decipher. This type of equipment is especially useful for individual quality control in manufacturing—for example in the testing of the finish of balls for ball-bearings, as seen in Fig. 48.

A third optical apparatus is the light section microscope, again developed by Zeiss. In this instrument a beam of light is projected obliquely onto the test surface through a narrow slit; at the surface it will cast a line of light (varying with the

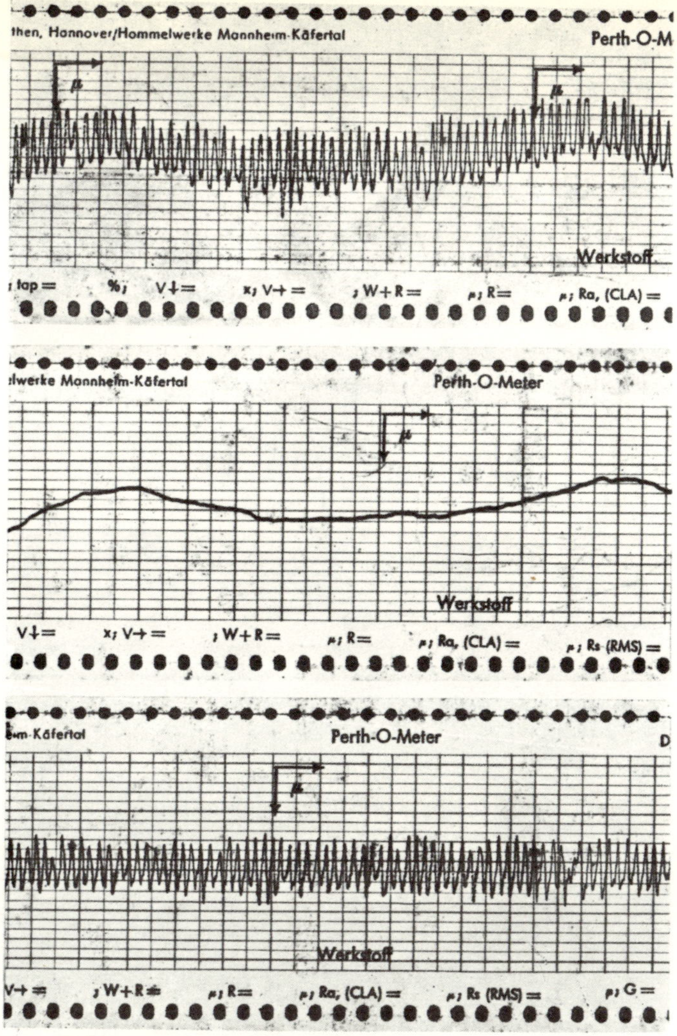

Fig. 42.—Surface profiles recorded with the Perth-O-Meter from a turned steel workpiece. Vertical magnification=4000×, horizontal magnification=40×.
Top: form defect plus roughness defect.
Centre: roughness subtracted, leaving form defect alone.
Bottom: form defect subtracted, leaving roughness alone.

Fig. 43.—Steel surface rubbed with emery cloth and photographed by oblique lighting. Magnification=200× (Rieter, Winterthur).

shape of the surface) which can be viewed and photographed through a microscope. The equipment is seen in Fig. 49, and Fig. 50 reproduces a few recordings made with it. The accuracy with which the surface is recorded depends very largely on the width of the light beam used, as well as on the sharpness of cut-off at the edge. Very fine irregularities are admittedly lost in the width of the line of light, but are still detectable by the shape of its edges.

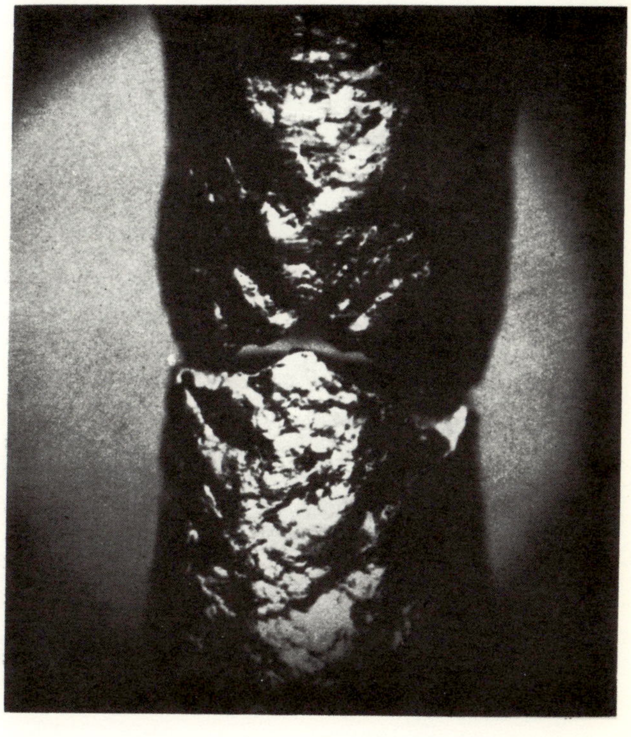

Fig. 44.—Surface of aluminium rod distorted and broken by tensile and varying pressure loads (Rieter, Winterthur).

Fig. 45.—Interference microscope by C. Zeiss.

The study of surface finishes is a comparatively recent development in the history of mechanical engineering but a start has been made to make up for lost time. A great deal has already been done, particularly in the development of ultra-sensitive test instruments, yet there is an impression that we can still see only the broad outlines of the enormous task this problem presents for technical development. A final solution still lies far in the future.

Surface position

Now that we have seen something of the difficulties that are involved in identifying the form of a surface, it will also be clear that it is no easy matter to say exactly where the surface is. What should we take as actually being the surface—an envelope profile, the highest of the burrs and irregularities, or some kind of arithmetical or geometrical mean value? We may be interested in any one of these, depending on the individual circumstances. If, for instance, we are dealing with friction bearings and are quoting a certain clearance which has to be observed, then the figure $1/h$ (h=width of the gap) will have a decisive influence on the shear stress in oil and the buoyancy effect, and we should be interested in the average value for $1/h$ at any point on the bearing surface. Today we are still rather lost when faced with problems of this kind, for we might say that without exception we measure the position of a surface at the tips of the peaks of its irregularities. Our normal measuring instruments—vernier calipers, gap gauges, inside micrometers, snap gauges and the like—all have relatively broad contact surfaces. If, for example, the three narrowest points in a bore touch the cylindrical surface of a plug gauge, then this hole has to be passed as satisfactory if this is the only check,

Fig. 46.—Bearing surface of a ring damaged by corrosion fatigue and seen by oblique lighting. Magnification=200× (Rieter, Winterthur).

Fig. 47.—The same surface with interference lighting, to show up surface topography.

irrespective of how often it departs from the proper circular shape in between these three points. For making a shrink connection, say, a check of this kind is next to worthless—individual projections on the surface will simply be crushed under the shrinkage stress, so that a friction contact is set up between shaft and hub, which will be quite unpredictable beforehand. Once, when I was trying to damp a very unstable governor by means of an oil film, a practical test showed that the design calculations had gone awry because the roughness of the surface involved was giving a much greater average clearance than had been allowed for. The surface was here quite clearly in a different place from that assumed for the purpose of the design calculations.

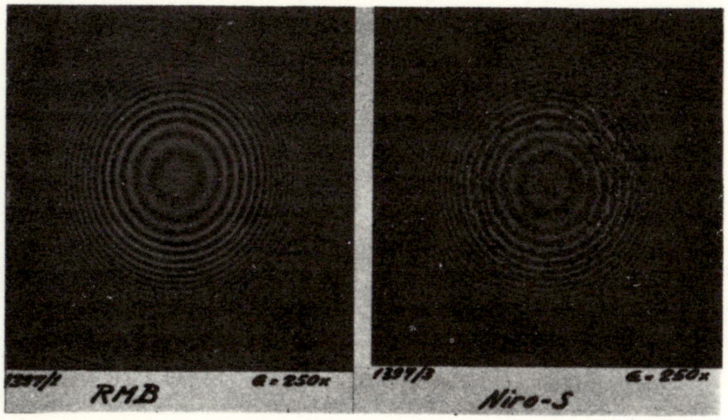

Fig. 48.—Surfaces of steel balls seen under the interference microscope. Left: in good condition. Right: with gross roughness (Paillard, Yverdon).

Interchangeability

Interchangeability—a concept which represents a major feature of modern engineering practice—is not enough. Two items may be interchangeable, and yet behave quite differently in actual operation. And it says nothing for the precision of a piece of machinery that the individual parts are within the prescribed tolerances, since the highspots on the surface may be crushed away as the parts are screwed tight during assembly, or may be sheared off during the first few seconds of test running. And yet despite this the whole of our present-day system of tolerances and fits is based on this highly-dubious way of determining the position of a surface. I cannot help feeling that far too little emphasis is placed on this weak point when the systems of tolerances are explained to students.

The starting point in creating these systems was the fact that it is never possible to keep the position of a surface exactly to the nominal figure. If you make a hundred workpieces of a given width w, you will find, inevitably, that there are certain variations from the prescribed value; and so it is obvious that the machine operators must be given a certain latitude in working to the nominal width w. However, in order to be sure that the item produced will still do what is wanted of it, this latitude must be specified. Consequently quoting a correct dimension means stating two figures, and not one—the upper and lower limits w'' and w' within which the dimension actually achieved w^* has to lie. The range from w'' to w' is called the *tolerance*, the figure w is the *nominal dimension* and the dimension that results in manufacture, w^*, is the *actual dimension.* It is not normal practice in quoting sizes to indicate w'' and w', but rather to give the nominal dimension and the range between this and the upper and lower limits, i.e. $w''-w$ and $w'-w$. We indicate, for instance, not the figures 49·9 and 50·05, but $50\,^{+0\cdot05}_{-0\cdot1}$. When using the metric system we always understand these figures as being in millimetres. The values $w''-w$ and $w'-w$ which are given in place of the upper and lower limits are known as *plus* and *minus* tolerances.

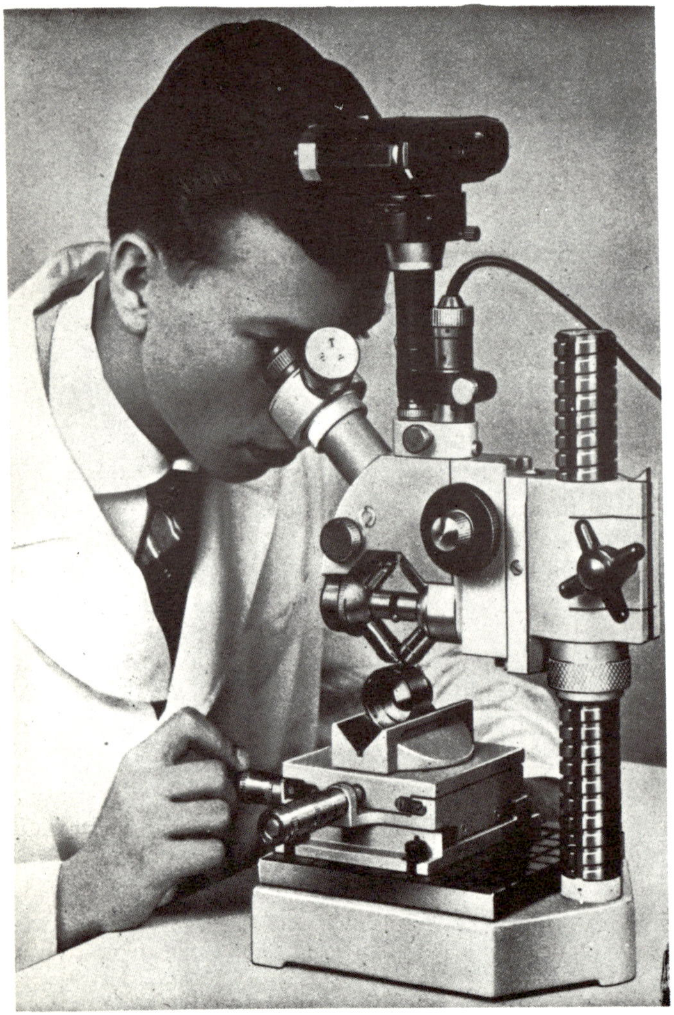

Fig. 49.—Light section microscope by C. Zeiss.

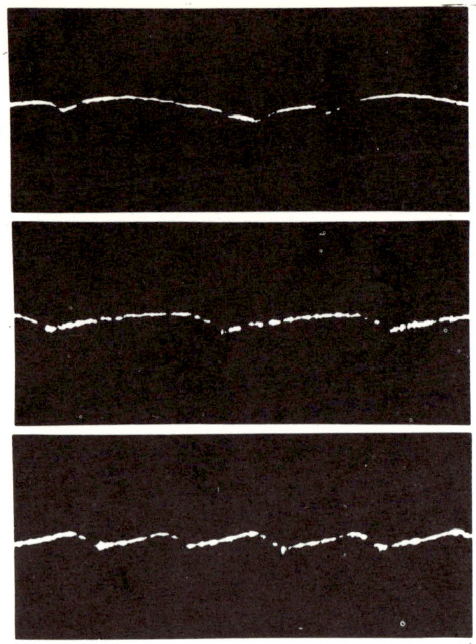

Fig. 50.—Light section recordings.
Top: filed brass, roughness depth = 32 μm.
Centre: faced brass, roughness depth = 20 μm.
Bottom: machined with carbide tool, roughness depth = 25 μm (C. Zeiss).

A check on the way the dimensions are being maintained in the workshop can be made with vernier gauges, but it is far easier and more reliable to make use of pairs of gauges which are made to precise limit dimensions. One of the pair is used for the plus tolerance and the other for the minus tolerance, and they are found especially useful for checking rotating parts. For holes we use *plug gauges*, while *snap gauges* are used for shafts (Fig. 51). The workpiece is within the prescribed tolerance, and is satisfactory, when one of the gauges will go through the hole (or across the shaft) but not the other; if both gauges 'go', then the workpiece is outside the tolerance limits and is rejected. These gauges are thus said to have a GO side and a NO GO side. It is, naturally, possible to make the workpiece pass through the jaws of a snap gauge if force is used—the proper way of using a gauge is for it to 'go' through by its own weight alone. If the NO GO side of a plug gauge drops through a hole when it is released, then the hole is too large and the job is rejected.

Operating such an inspection system does, of course, mean substantial costs, since the limit gauges have to be made with extreme care and accuracy. So as to keep costs down to a minimum it has been expedient to apply the gauge system in as many firms as possible; this is all the more advisable since it also allows the interchangeability of components. The

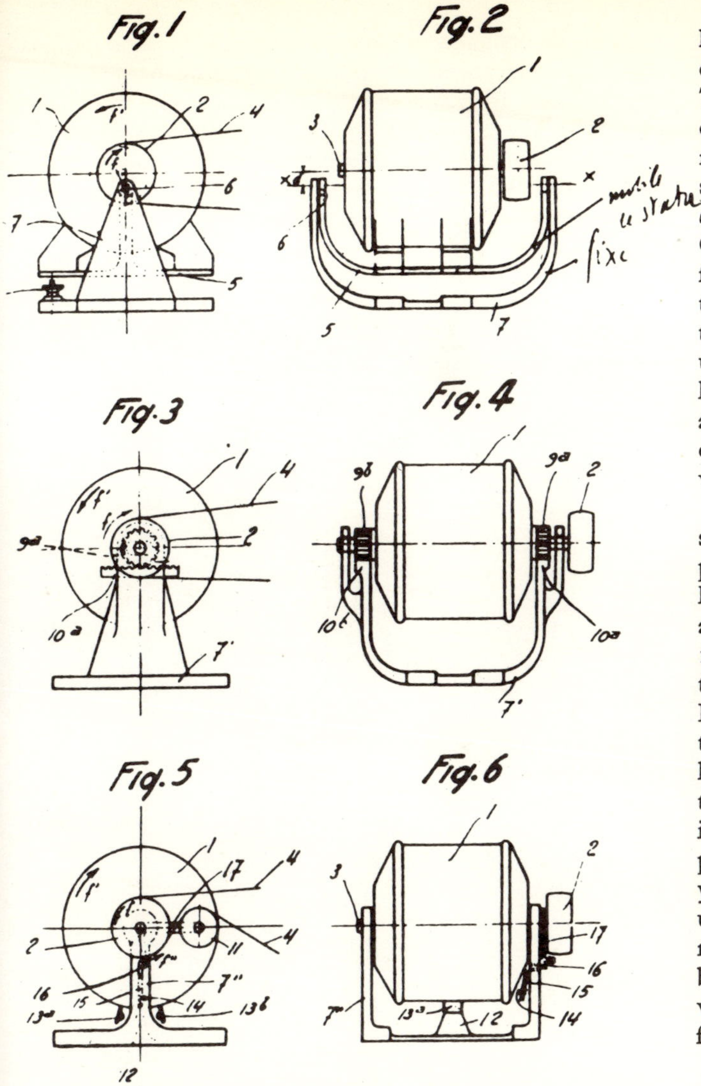

Fig. 3.—Illustration from a French patent application of 1935 on a self-tensioning belt drive system. These sketches are bare representations of principle, with not the slightest similarity to design drawings. Since the inventor failed to develop his ideas into actual designs, no such drive system was ever put into use.

command. It means, however, that the student with scientific and technical training must learn how to design and that the designer must extend his studies as much as possible in breadth and depth. It would be very beneficial if this simple formula were given more attention in the training of the next generation of engineers.

It is important to stress this relationship of invention to design, since many people—especially young engineers—who have not yet grasped the importance and scope of design work often devote themselves entirely to the work of invention. They have a constant fund of ideas on what still needs to be done, and they hardly pause to reflect that the design work required to bring the average invention to the production stage involves several thousand hours of work and can swallow up tens of thousands of pounds in development costs. Contrary to what most people think, invention comes about fairly easily, and often the need is more to curb the inventor than to encourage him. 'Less invention and more design' was the pithy maxim which one former lecturer used to impress upon his students. Inventions pure and simple make very little impression on the experienced engineer—to the disappointment of many an inventor; what can make him quickly drop his reserve, however, is a working prototype—in other words, a complete design.

In this connection it is in some ways surprising that society should have created a broad-based and costly institution—the patent system—to protect the inventor. An inventor can seek legal protection for his brainwave and society guarantees him a monopoly on the idea of his invention for a period of 15 or 18 years. He alone is entitled to make use of it even though thousands of others may have the same brainwave after him. No wonder that every idea, no matter how half-baked it is, is trotted along to the patent offices and a claim for protection lodged, until these institutions hardly know how to cope with the flood of applications. If we examine the fate of these inventions from the date of the application we find that most patent claims are allowed to lapse after a few years. As the years pass, the number kept up become smaller and smaller, until at the end of the period of protection only a handful remain (Fig. 5). It can be assumed that these survivors have been successfully put into practice; all the others represent wastage—a fact that ought to make the inventor think carefully and exercise restraint.

Many inventors do not realize that even the best of inventions requires further design work if it is to find practical applications. This explains why even worth-while inventions often fail to get past the paper stage, and why so many inventors never see the rewards they hope for. With a few notable exceptions, inventing is bad business, both for the inventor himself and for the State which has to maintain a complicated apparatus for protecting what are for the most part doubtful interests.

The designer, too, as well as the inventor, is naturally interested in having protection for the results of his work. If a successful design gets into the hands of a competitor, it is usually a serious blow for the firm in question. If we remember that the best engineering skill and experience in a firm may be embodied in a design for, say, a jet engine or a modern packaging machine, we can appreciate the need for comprehensive protection for this kind of work just as much as for that of the inventor. Yet the designer does not enjoy protection—at least, not to the extent provided for the inventor. There is a law governing unfair competition which is cited by lawyers whenever a case arises; but this does not specifically mention design as such, and so the designer often loses his case to whoever has copied his work. The latter also has on his side the fact that the highest courts in the country[1] have declared design to be common property, i.e. in their opinion designs can be copied by anyone. In practice, therefore, designs enjoy no protection. Everyone in engineering and industrial circles is well aware of this state of affairs, and an attempt is made to erect some sort of protective barrier by maintaining strict secrecy or keeping the design drawings under lock and key. Firms also try, as far as possible, to extend the protection given to inventions to cover designs as well, by registering designs in the guise of inventions; this leads to what are known as

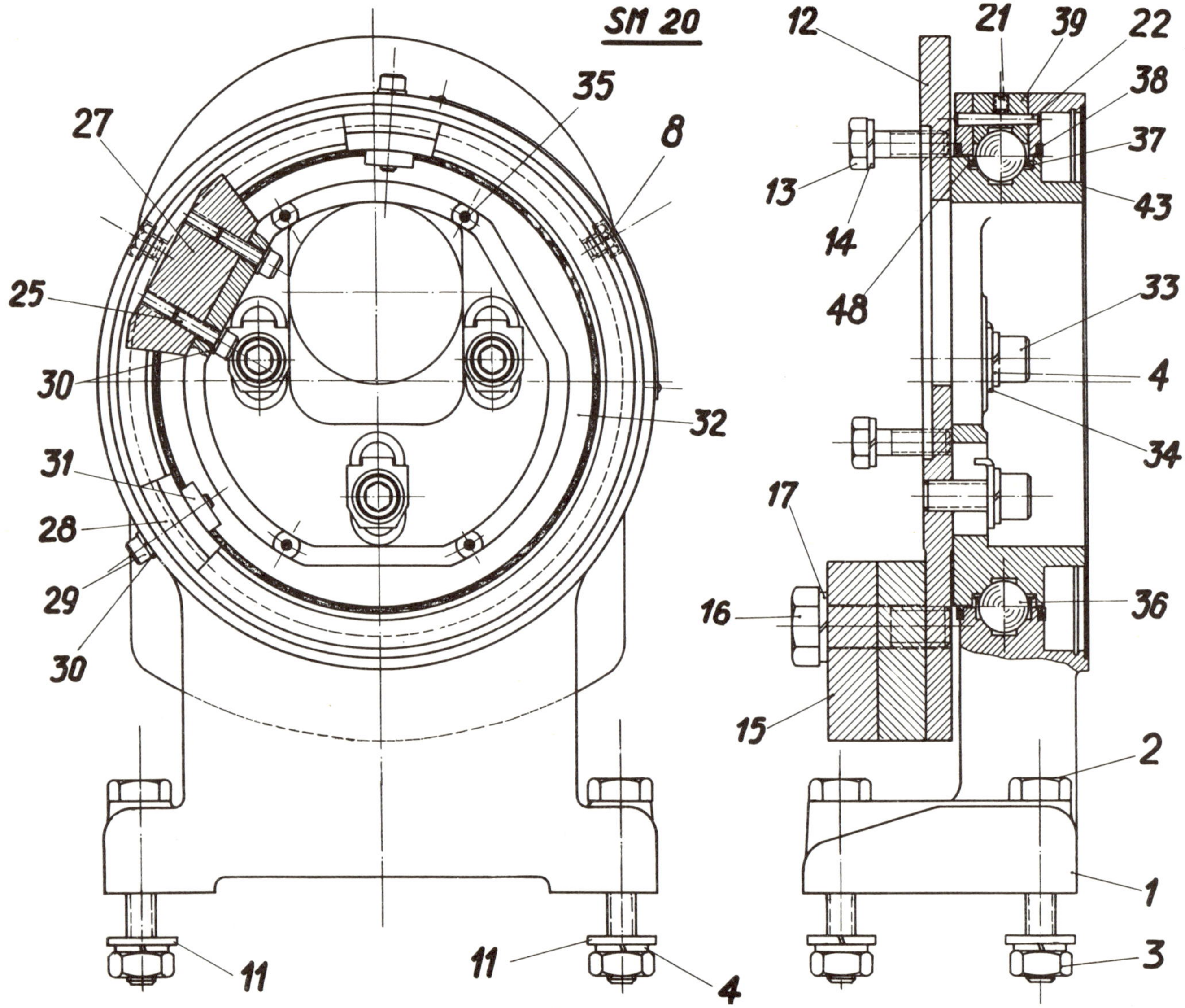

Fig. 4.—Design for the motor suspension of a self-tensioning drive system. Not until detailed designs like this were produced did it become possible to put this system of drive into use (Antriebe AG, Rapperswill).

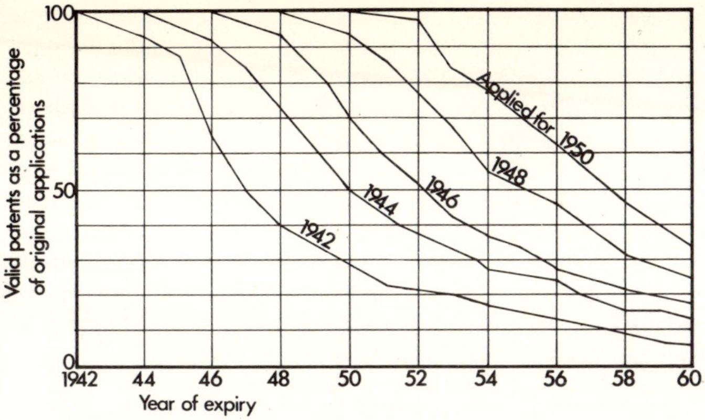

Fig. 5.—Lapsing of major patents applied for at the Berne patents office (according to the Swiss Annual Index of Patents on Inventions and Registration of Samples and Models). It can be seen that the majority of patents registered have been allowed to lapse after a few years.

'design patents'. However, none of these measures gives any really reliable protection so that there are always quarrels about the pirating of designs. The courts display a marked lack of certainty in their judgments, doubtless because of the contradiction between what is felt to be just and the defective legal propositions[2].

Design components

As we have seen, design is invariably the further development of one or more inventive ideas (old or recent). In developing these ideas, the designer makes use of a whole range of constructional elements such as screws, wedges, levers, wheels and so on, which date from the very beginnings of technology and were themselves, at one time, inventions. As fresh inventions are constantly being made, the number of such constructional elements is continually growing; when they reach the stage of being common knowledge, and when the patent rights on the invention have expired, they are known as *machine elements*. Any mechanical device invented now will become—if it is useful—a machine element after a few years.

A designer should continually endeavour to increase the store of such constructional components at his command and should keep an eye on specialist journals and, particularly, current patent applications. The larger and more varied his stock of such design components, the more productive his work will be, and the more elegant and sophisticated his designs will become. A constant effort to extend this knowledge is necessary since one's store tends to shrink as items are forgotten and cease to be used. Designs then become flat and monotonous since the same components are used again and again, and finally it becomes impossible to tackle difficult projects at all. Every experienced designer can confirm that this danger really exists, and it is not taking things too far to recommend that the designer makes lists of those constructional resources he has at his disposal.

Standard specifications are a first step in this direction; they set out a large number of the common constructional components such as bolts, cotter pins, tube couplings, fittings and so on. Only occasionally does Industry itself take steps to inform designers, by bulletins or lectures, about the latest design components, but this is something that ought to be widely fostered. There is a whole range of machine elements which we know to be good and which show no disadvantages; yet they have not become popular among designers. The only possible explanation is that the ways of informing designers about such useful innovations have not been sufficiently developed. This is perhaps the case with the cone locking ring (Fig. 6).

The need for a clear statement of the problem

The success or failure of a design may often be decided in advance by the way in which the design problem is presented. Many a design has failed to make its mark because the specification was wrongly prepared. Any major design project should be preceded by a careful assessment of what is required; as well as asking 'What is wanted?' we must also ask 'What can be specified in the given circumstances?'

There can be widely different intentions behind one design, and they may result in equally varied styles of design. An assembly jig which is to be used only once will not be made in the same way as one which will be used repeatedly over a period of years. The kind of personnel available for operating a machine—trained mechanics or unskilled workers—will also play a part in the design. The place where a machine is to be set up—in a closed machine shop under ideal conditions, or out in the open where it will be exposed to heat, cold, rain and snow—will again make a difference. The main requirements for a pump which is to deal with ground water in a pit will be rugged construction rather than high mechanical efficiency. There was an occasion when a bore-hole pump of this kind fell down among rocks while being transported across a ravine and it was expected to be unusable or at least in need of major repairs; yet it immediately started working again when it was connected to the mains. This is the kind of pump

1 Although this passage refers to the situation in Switzerland, the position is similar in other countries.

2 Further remarks on this subject can be found in an article 'Legal rights on technical designs against identical copies' in *Schweiz. Bauzeitung*, 1953, No. 48 and *Schweiz. Juristenzeitung*, 1953, Vol. 24.

required for that sort of job and a comparatively low efficiency is not a prime disadvantage. Specifications must take account of such considerations.

Some 20 years ago when the first gas-turbine locomotive was built, the designers adopted, without giving much thought to the problem, the control system that had been used for similar stationary installations for many years, i.e. hydraulic controls. Initial tests with the system were completely satisfactory. But when the locomotive had stood in the open all night in a temperature of −20°C the hydraulic fluid was frozen solid and it was quite impossible the next morning to run up the turbine. The entire control system had to be replaced by an electric one. This design had failed because the problem had not been properly set out. Allowance should have been made for the fact that such equipment has to work in a wide range of ambient temperatures, and the specification should have stipulated that it must work properly at low temperatures.

This is one example among many. It shows that the success of a design project must be considered from the outset—at the time the specification is prepared. The first problem is that of deciding who is competent to specify what is wanted in the design. Usually a new design is initiated by the client or agent—they are the first people to be aware of the need for something new. But this is no reason for assuming that they are capable of identifying exactly what is needed or are capable of writing a specification. These tasks call for expert and detailed knowledge which will normally be available only to the person responsible for directing work, i.e. the chief designer. The design job will be done best if he assumes responsibility for saying what is to be done. He is at the centre through which all information must pass, and it is he who will analyse the requirements and make the final decisions.

The right amount of perseverance

As he goes deeper into the whole subject of design, the beginner will probably be surprised by the strange laws which govern the way it works. When first seeking the solution of a design problem, he can see nothing but difficulties; the conditions to be met appear so contradictory that it seems to him unlikely that they will ever be reconciled. At this stage the designer is tempted to give in to the difficulties and to delete one item after another from his list of requirements. It is

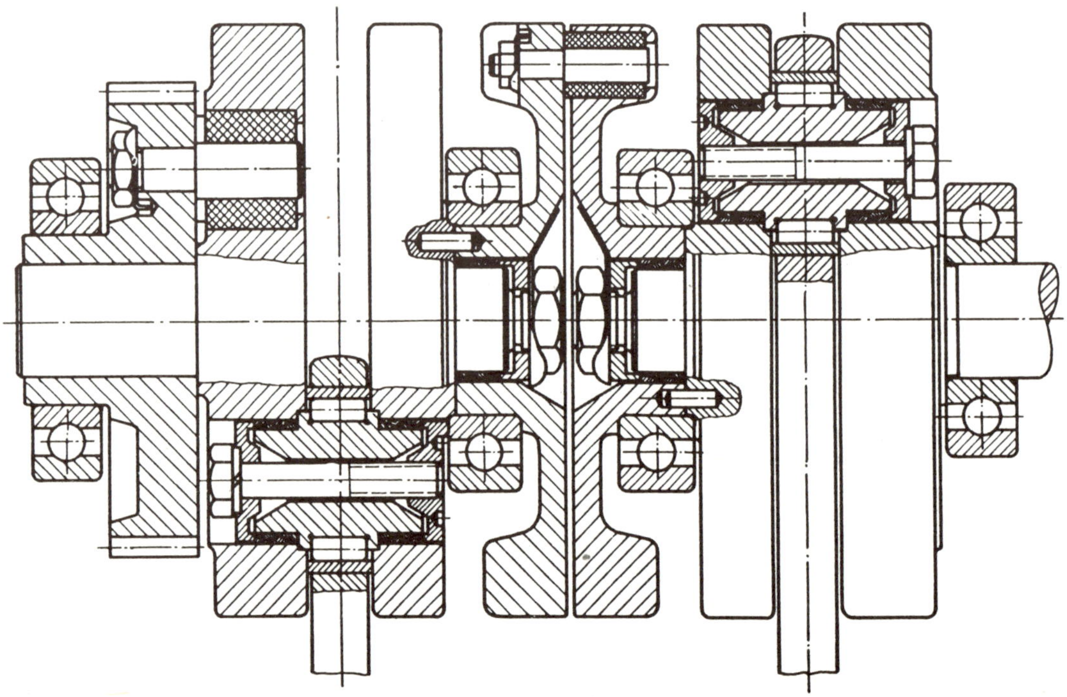

Fig. 6.—Tapered clamping rings used as a means of joining a shaft and a hub offer innumerable possibilities in mechanical engineering. Here we have an example of them being used to assemble a crankshaft from separate units; this system of joining the parts is just as good as a shrunk joint and, unlike the latter, can be taken apart again. (*Photo: Waggonfabrik Uerdingen. Rheinland.*)

astonishing how quickly he may let himself be persuaded into this sort of surrender, and how he may rapidly reduce his demands until finally the whole project is not really worth undertaking. Unless, in these circumstances, he pursues the final aim with perseverance, he will never become a successful designer. It is almost as if Nature were testing our resolve, for she only gives way when we have made up our minds to be bold. This attitude is shared by all good designers and seems to be a *sine qua non* for successful design work.

Once the designer has adopted this approach, however, obstacles begin to crumble and avenues open up where previously there seemed to be only an impenetrable wall. He comes to realize that circumstances have an unexpected tendency to adapt, and that in the last resort Nature is always prepared to yield if we make a big enough effort. Things go well so long as we go about them in roughly the right way, and do not defy logic and good sense. Often, as we work on a problem, inspiration comes more readily and ideas follow one another almost without effort.

Sometimes it is a matter of deciding when to be satisfied with the result and allow the design to go into production. Every successful designer is striving for perfection—consciously or unconsciously—and this basic attitude is essential if anything worth while is to be achieved; but it is something that must be kept under control, for otherwise it will develop into a search for the philosopher's stone. One meets the designer who always wants to make just one more improvement, and can never make up his mind to call a halt. If a designer is always trying to produce something yet more perfect, all he will succeed in doing is bankrupting his firm. He must realize that there is a limit to his capabilities and come to terms with it. He must appreciate that fundamentally there is only a difference in degree between Stone Age man and ourselves. If we look at modern machine components, we can see that they are all, without exception, extremely simple; and this is true of all our technical creations.

Fascinated though we may be at the beauty and elegance of a modern jet airliner, we may see, if we look at it more closely, that the sheets of aluminium have seemingly been riveted piecemeal to the spars and frames, and have only been moulded into a suitable shape with some difficulty. We may be dazzled by the chromework on a car—but a glance at the engine or chassis is enough to make us realize how much effort has gone into it. We find the same thing when we consider the functional efficiency of our technical achievements. We put all the skill at our command into building a piece of machinery; yet when we use it we find that despite all our efforts it has many shortcomings and is far from being able to do all we would like of it. Try asking a man who has had a leg amputated—however technically perfect his artificial limb may be—whether he would not rather have his own leg back! Whatever we do is inherently marked by imperfections which are only just hidden from view; whatever we construct is no more than an improved kind of crutch to help us get through life. The conclusion we must draw is that although we have a basic mastery over Nature, it is a mastery within limits; and the more we try to improve our achievements and approach perfection, the more resistance she offers.

Nevertheless we are irresistibly committed to this striving for perfection and cannot help continuing our search for it. We have within us a unique concept and standard of perfection, and where this standard is not provided for us by Nature, we reach it by a process of ingenious abstraction, as in the Carnot thermodynamic cycle. We not only devise this standard of perfection but we set it against everything we have created. We feel unable to put it on one side as something that does not apply to us. The most eloquent example of this lies in the notion, which we ourselves have evolved, of mechanical efficiency; this amounts to measuring our achievements against the standard of perfection.

We are continually improving our performance—and yet we never achieve perfection. So the designer must realize that he cannot make unlimited changes and improvements to his design. He must learn to recognize the point at which his efforts must be called off, and a start made on production. Even the best products of the human mind have some ambitions—even feasible ones—unfulfilled. Everything created by Man, however good it may be, has room for improvement.

Design and calculation

Form design—deciding the shapes of the individual components of a machine—is the fundamental part of the design process. It represents the truly creative stage—the creation of a mechanism or structure in the mind's eye using intuition, thought and imagination. The designer tries to reach the best possible concept without, initially, giving any thought to the relationship of individual dimensions—perhaps with no particular scale in mind. Successful form design is an art and is thus particularly difficult to systematize; yet a great deal can be achieved by improving one's system. When the beginner fails, it is usually due to an unsystematic approach. He should never attempt to produce shapes before he knows their real purpose. Many designers try to put the cart before the horse, ignoring the fact that every design must dovetail in some way with something which already exists. This should always be clear from the design specification, in which the designer has a foothold from which he can proceed to develop his design. The specification will always identify a part of the mechanism which is the focal point of its function. In a riveting machine, for instance, it will be the point at which the pressure tool compresses and shapes the rivet; in a shaft coupling it will occur where one part engages the other; and in a reciprocating compressor it will be the face of the piston which displaces and compresses the gas. It is always best to start the design from this focal point; then, step by step, and making new footholds one after another, the design is developed in one or more directions at a time—both figuratively and in three-dimensional terms.

In this exacting work calculation is never more than a tool. It will help the designer enormously in deciding dimensions and forms and sometimes it will carry him a long way; but he will always return to his preoccupation with form design which is the thing on which the machine depends. Although numerical work plays an important part, and is indeed indispensable to the whole design, it is completely non-creative. It is a tool, and only the person wielding the tool does the creative work. The expectation cherished (rather rashly) in some quarters that the electronic computer would largely replace human intuition has not been fulfilled. A design can never start with calculations; imagination always comes first.

The eye plays an important part in form design. It notes conditions and relationships unconsciously and guides the designer's hand in drawing his lines, sometimes leading him to solutions he would never have discovered by conscious thought. Experience shows that the eye can often take the place of calculation. The practised eye will notice, for example, if the individual links in a chain are of differing thicknesses, and the weakest is found without need for calculation. The same applies to the various parts of a design. Time and time again we find designs where, without a great deal of thought, we can point to the spot where any fracture will occur. In such a case, the design is not ready to be put into production. Great care must be taken to ensure that all the parts of a design offer the same resistance to failure. Although this principle is obvious, it is ignored by many designers.

This principle of uniform strength is particularly relevant where exact data on working conditions are out of the question, and consequently it is not possible to tell in advance the stresses to be borne by the individual parts of a machine. How, for instance, can we find out what stresses will be put on the jaws of a rock crusher when they close on a particular hard piece of stone? With a machine of this kind there is no alternative to giving all parts the same degree of strength (and then there is the question of the overall reliability against failure). It is the special function of the human eye to ensure that this principle of uniform strength is followed, and there are many designers who have brought this ability to a high degree of perfection. They set out dimensions with a decisiveness which is as confident as if they had been obtained by calculation. This ability is sometimes called 'technical feel' and anyone ignorant of the mental processes involved is inclined to look upon it as a natural gift, a sixth sense. Basically this 'technical feel' is no more than making comparisons between similar circumstances or, in other words, the use of a 'mental yardstick' developed by long experience. Without a system of reference, 'technical feel' cannot make any sort of accurate judgment. Even a skilled engineer, depending solely on his 'technical feel' would be unable to say that the greatest shear stress in a 50 mm (2 in.) shaft is 10^8N/m^2 (15,000 lbf/in.^{2}) when the torque is 2500N m (2000 lbf ft.). There have been cases where designers have been several hundred per cent out in their estimates without the slightest feeling of doubt in their mind. We must always expect such errors when there is no opportunity for comparison on which to base the judgment. To avoid nasty shocks calculations must be carried out in all cases of doubt: however uncertain the assumptions behind these calculations, they will always be more reliable than the indefinable 'technical feel'.

The art of moderation

What I always admire in past designs is the very evident art of moderation that has gone into them. A sense of balance and proportion is the most striking feature of any proved design. Everything is well ordered, equally removed from niggardly penny-pinching and extravagance. The experienced designer is not willingly pushed to extremes in any direction; the 'golden mean' is usually most successful in the end.

In designing we try to economize on materials, and we also keep the number of separate parts to a minimum; a steady reduction in the amount of material used is a special feature of recent developments in design and today's machines are smaller and lighter than those of the past for the same output. The Swiss engineering industry has no raw materials of its own; every piece of iron has to be imported, so that the economical use of materials is specially important to firms in Switzerland. It must be remembered that any move to save material will nearly always result in a more complicated shape for the component concerned, and will consequently mean a greater outlay in labour. Material costs and labour costs tend to be inversely proportional to each other, and the optimum point for any design will occur where the sum of the two is at a minimum (Fig. 7). But this is not the only criterion for deciding how to go about building a machine—a proper decision depends on the overall technical and economic circumstances affecting each individual design. With design projects typical of general engineering, this usually means a form of construction which is heavy and bulky rather than elegant (Figs. 8 and 9). The beginner should keep to examples

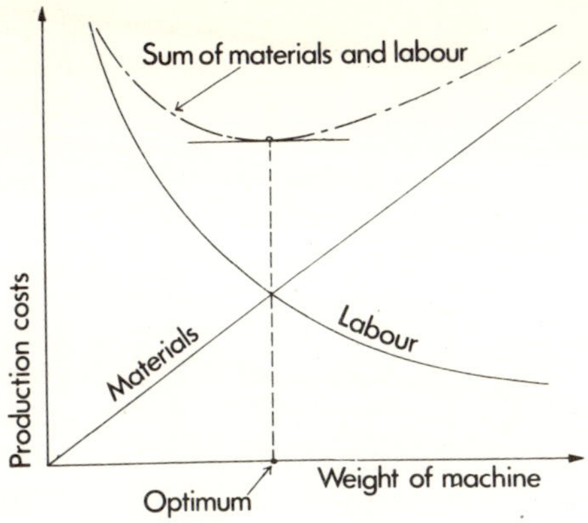

Fig. 7.—Calculation of optimum weight of a machine from cost of materials and labour.

of this kind, since they exhibit most clearly the principles of design which can be applied universally.

In trying to design economically the designer is constantly faced with the problem of whether to combine two or more functions in one component, a possibility suggested by examples in Nature. In an engineering context, this may involve a decision on whether, in a power plant, to control and cut off the flow to a water or steam turbine by means of a single valve unit, or whether to split the two functions and have a separate control valve and cut-off valve. If we examine the practical solutions adopted by engineers, we find that nearly always the answer is to separate the functions. It is seldom satisfactory to combine several parts in one in an attempt to make things simpler and cheaper; such multi-purpose components make but a poor job of the various tasks they have to cope with. Here, too, it seems that no man can serve two masters and get away with it.

Only in rare cases does such an approach come off. An example which should be mentioned is the common combination of axial and radial bearings, seen in Fig. 10; but the experienced designer will only do this sort of thing as an exception. His normal approach is to have each separate function handled by a separate component designed for the job, and to ensure also that this component is independent as regards fitting, servicing and dismantling. The beginner may consider this complicated and unnecessarily cautious; but watching a machine being produced, assembled and serviced, and having personal experience of these things, will soon convince him otherwise.

It is also a mistake, when designing, to try and make a saving on the small and apparently unimportant parts. The designer must learn at the outset that the distinction between 'main' and 'subsidiary' has no meaning in machine design—there are only 'necessary' and 'superfluous', and the latter will always be omitted in any case. It may surprise the young designer to be told that every component in a machine should be accorded the same status; he feels that alongside the major items there must surely be minor 'bits and pieces', the design of which requires very little attention. The designer must get rid of this idea from the very beginning of his career.

The lubrication is just as important as the bearing (Fig. 10), the waste-oil drainway is just as important as the gland itself (Fig. 11) and the adjusting mechanism is just as important as

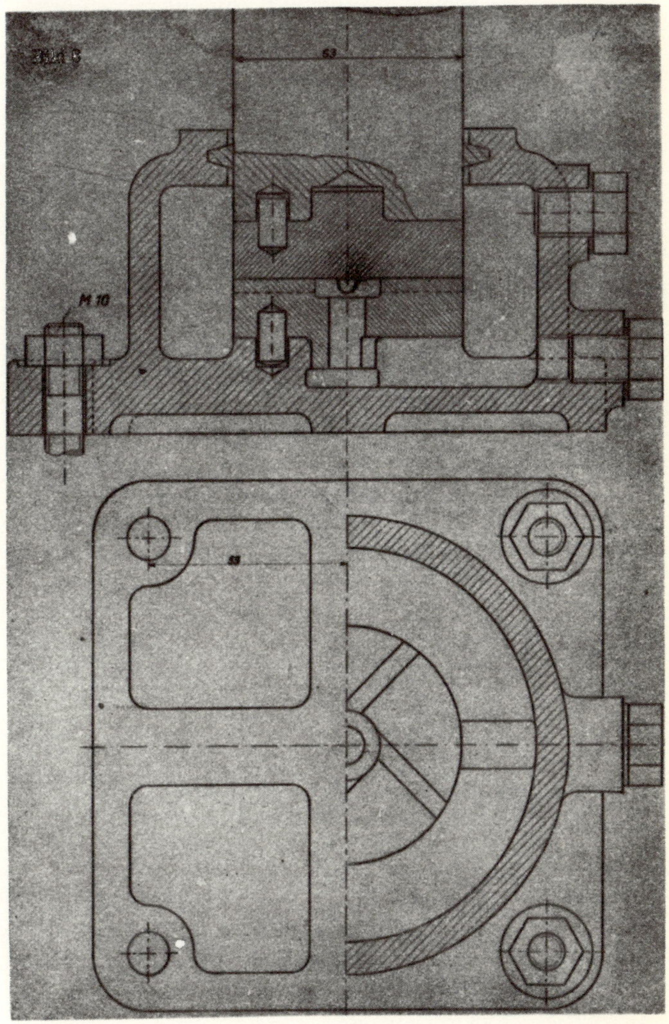

Fig. 8.—Example of general mechanical engineering design; a thrust bearing. Note the relatively heavy and simple form of construction.

the brakes (Fig. 12). The list of examples could be continued indefinitely—it covers the whole field of engineering. Long experience of steam turbine operation has shown that 80 per cent of all breakdowns can be traced back to ignoring this rule and we can expect a similar state of affairs in other fields. Time and time again designers give too little attention to details. When newly-developed designs fail during test runs, it is nearly always the most commonplace components—bearings, couplings, gaskets and the like—which are to blame and not the principle of the machine in itself. When, for example, the chairman of a company asked his departmental manager why a certain new design was not going ahead despite the expenditure of tens of thousands of pounds, he got the astonishing reply that it was the ordinary machine parts that were constantly breaking down and causing delays. In another case, a customer refused to accept a heating plant costing £200,000 because of the chronic failure of a perfectly ordinary sealing ring. It only needs one part to fail for an entire installation to be put out of action.

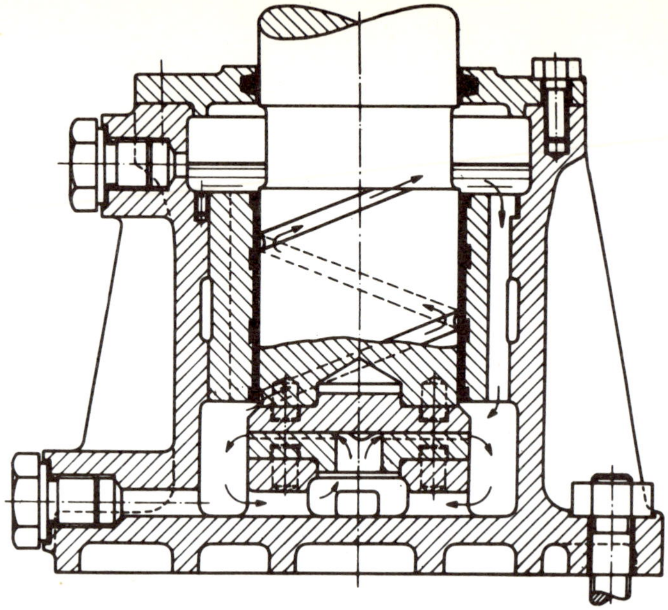

Fig. 10.—Friction bearing for radial and axial loads, with special means of supplying oil to both bearing surfaces.

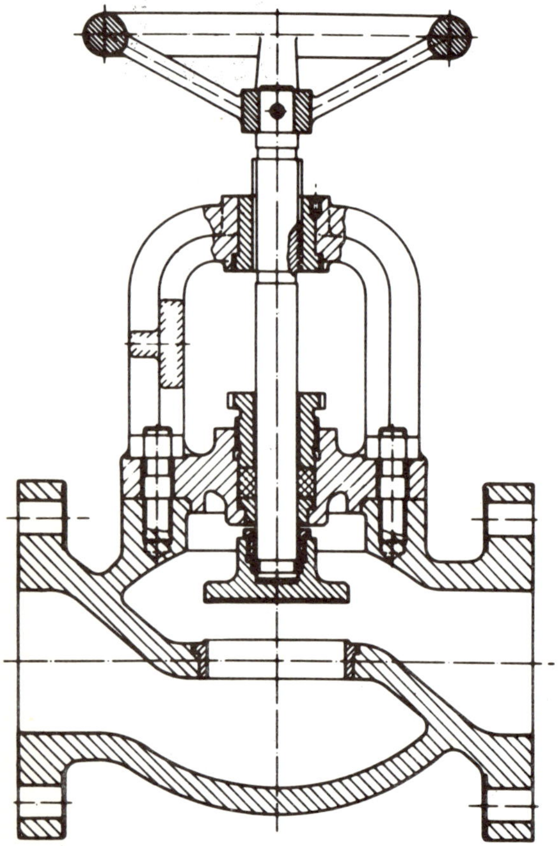

Fig. 9.—Example of general mechanical engineering design. Note the relatively heavy and simple form of construction.

As a consequence it is an inexorable rule of engineering construction that every detail must be dealt with thoroughly, and not left to chance. The designer cannot offer the excuse that, because of pressure of work, he did not have time to see to this or that; unless his design has covered every detail his work is not finished, and he can be sure that anything he thought he could ignore will be wrong and give trouble. The more consistently a designer directs his attention to details, and the harder he tries to attain a tidy solution, the quicker the project will proceed. How he is to cope with a thousand-and-one details without getting lost among them is a matter of working with a method.

Many designs are notable for doing one and the same thing in several different ways—a hub may be shrunk on to a shaft and then keyed as an additional safeguard to prevent slipping, or two parts may be riveted together and then, in addition, welded along the edges. Sometimes this doubling-up may be for a good reason; if the failure of one particular solution to the design problem could lead to a serious accident, for instance, then an additional solution is usually adopted. For example, every boiler has two safety valves and every car has two braking systems; if one fails, the other is there to prevent disaster. But apart from these cases we come across designs in which this 'belt and braces' approach has been used despite the absence of any special risk. This invariably stems from lack

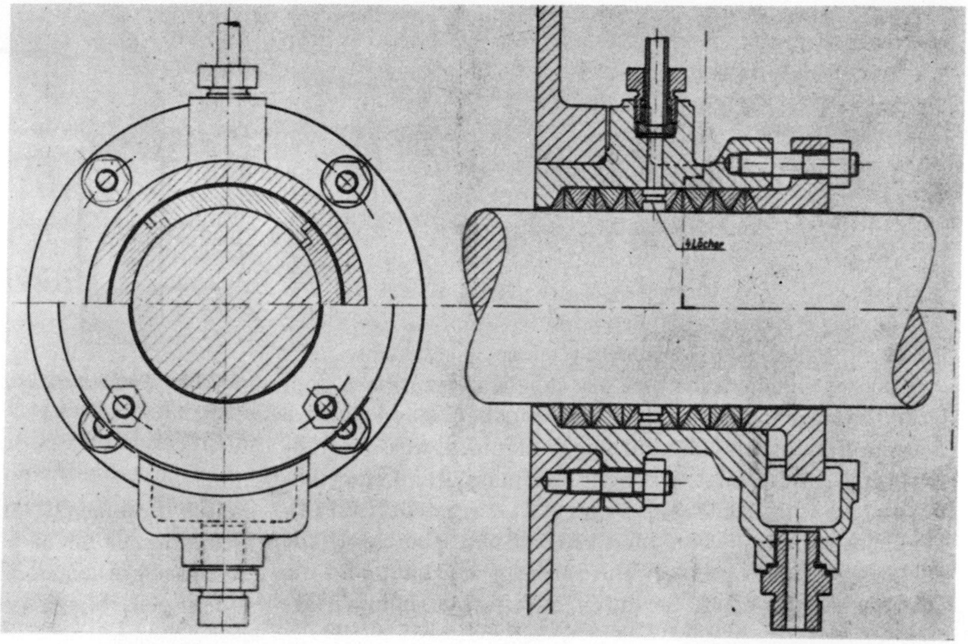

Fig. 11.—Gland with carefully-designed arrangements for draining seepage and supplying oil to the packing.

of sureness on the part of the designer. He does not trust his first solution and backs it up with another, and so on. Such a designer should find another occupation or drastically overhaul his working methods. The habit comes from stopping halfway and never following anything through to the end. One pays heavy penalties for half-measures in the design world—it is better to do nothing at all than to half-do something. There is no surer way of wrecking a business than to constantly add one stop-gap solution to another.

Design work can only be successful if the basic rule of 'one problem, one solution' is followed. If a hub is to be shrunk on a shaft, the whole thing must be carefully calculated, the

Fig. 13*a*.—Heavy crankshaft for a marine diesel engine, fabricated from separate webs and journals by shrinking. Although the parts are held only by friction, this is evidently adequate for coping with the heavy loading involved. (*Photo: Gebr. Sulzer, Winterthur.*)

Fig. 12.—Motor-car brake assembly.

Note the neat arrangement for easy adjustment (arrow).

amount of shrinkage accurately produced and the surfaces made fair and smooth (Fig. 13*a* and *b*). Similarly, if a gap is to be sealed the gasket must be carefully selected and properly

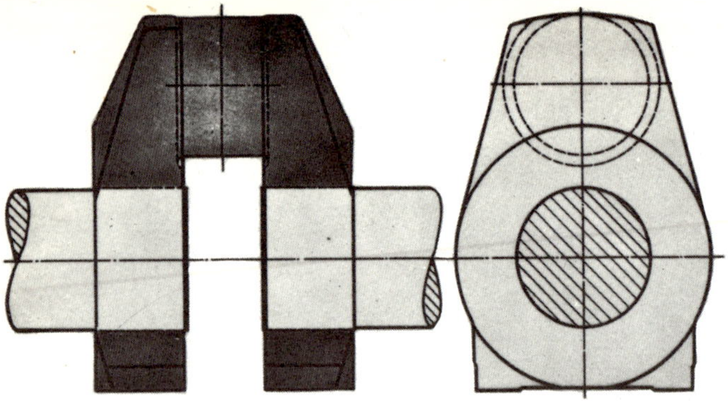

Fig. 13*b*.—Workshop drawing for the shaft seen in Fig. 13*a*; this shows that no shearpin or similar device is used for fastening the shrunk joints.

fitted, and the surfaces must be flush. Work carried out in this way ensures complete success and no additional precautions are necessary.

When, for reasons such as those given above, duplication is thought necessary, there will be legal requirements to be followed. In such cases the design is planned so that the systems operate independently of each other and, to ensure an effective reserve, care must be taken that one of the devices is called into use only if the other has failed. Where a double seal is used, the reserve gasket will remain unstressed so long as the first is functioning satisfactorily, so that premature loading does not reduce the full effectiveness of the reserve (see Fig. 14). Where there are no specifications of this kind, the designer can be sure that duplication is unnecessary and that a sufficient degree of safety and reliability is possible with a single component.

The courage to be exceptional

Apart from the 'golden mean' designs there are always others which are extreme in some way or other and where the design problem has verged on the impossible. In these cases the designer is forced to abandon his normal approach, the careful avoidance of every possible risk and the cautious treading on the firm ground of what is tried and proved (Fig. 15). Naturally, designs like this have features quite different from those seen in general run-of-the-mill engineering work; if a modern aero-engine is placed alongside a stationary petrol engine, almost everything will be different and no two components will be found that are identical.

Experience shows that it is so difficult for a designer to readjust his attitude to designs of this nature that it is often better to give these projects to entirely new and unprejudiced people. The specialist often finds it difficult to believe that there are any possibilities beyond his past experience and so many new and creative designs are the work of outsiders rather than of the firm's own experts.

Such designs are often the start of a fresh line of development and can be recognized as pioneering work (Figs. 16 and 17); the transition from the steam engine to the steam turbine, and from the latter to the gas turbine, was made in this way. Only courageous designers are capable of work like this, for they are following paths which offer no guarantee of success. There have been countless attempts at creating something fundamentally new that have come to nothing because of unexpected problems. Often the work has been taken over and carried further by someone else. Only a few designers are willing to contribute in this way, but the major part of all technical progress is due to them. Most of them have done more for the welfare of mankind than many of the generals and politicians whose monuments are put on pedestals around our streets. Whether to continue to gather a scant harvest from the soil of past deeds or to make an all-out effort to break new ground is a difficult choice, because the designer is not alone; behind him are many people whose livelihood depends, at least partly, on the outcome of his venture. This responsibility weighs all the more heavily on the designer since their support is not usually given freely but has to be gained by every means at his disposal. Most people prefer a safe journey

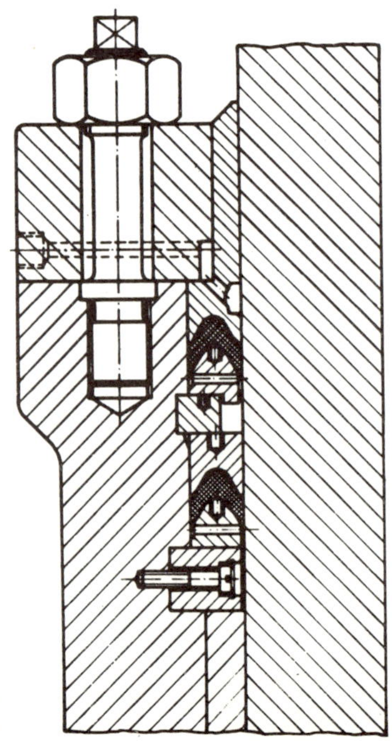

Fig. 14.—Double seal in a hydraulic press. The main seal (below) rests on a ring insert, putting no pressure on the reserve seal (top).

Fig. 15.—The idea of using the freight load as a load-bearing part of a railway truck led to this remarkable solution to a transport problem (generator made by Brown Boveri for the steam power plant at Weisweiler).

Fig. 16.—This is probably the most accurate horizontal boring machine made at the present time, with programmed control, produced by Messrs. Dixi of Le Locle. Driving the main spindle of this ultramodern machine by means of a flat belt, and mounting it in pedestal supports, was a major and difficult decision.

Fig. 17.—Drive side of the machine seen in Fig. 16, using nylon-reinforced flat belts (made by Leder & Co., Rapperswil).

along a well-worn road, and only one person in a thousand can inspire them to out-of-the-ordinary efforts. If the goal is reached, success is afterwards taken for granted; if not, the leader is discredited and cast aside.

The fact that this one person in a thousand is always to be found, ready to take on such risks and work for progress—although standing to gain very little and possibly to lose a lot—is due to a sense of vocation. Neither ingratitude nor lack of understanding will deter these pioneers of progress from their hazardous ventures. Their attitude contains a kind of quiet heroism which is very seldom spoken of, because it is usually unrecognized as such. All kinds of people are honoured, but seldom does a designer receive public recognition for work of this kind.

The risks involved in such speculative ventures could be reduced if more thought were given to whether the time were ripe for them. Special creative engineering of this kind calls for a correspondingly high level of general technology and, if the current technical resources are outstripped, it becomes impossible to carry through the new design. We now know, for instance, that early attempts (by Holzworth) to build a gas turbine were quite simply ahead of their time, and that the technical resources of the day could not cope with such a project. Such examples of designers being too far in advance of their time are by no means uncommon.

Design and aesthetics

What has machine design got to do with aesthetics? This question comes to mind whenever we look back at early examples of mechanical engineering, for technology became the enemy of beauty wherever it appeared. The characteristic lack of beauty in such things as cranes, trellis masts and dams was obvious; sometimes, indeed, they were downright ugly. The disfigurement of nature by technology was alarming and the gulf between the two seemed unbridgeable. Gradually, however, the dark clouds began to clear and it was realized that the ugliness found in technical creations was not intrinsic, but rather an expression of their imperfections and of the fact that the technical products of the day had not reached full development. The engineer had not found the complete technical solution to his problems and had often taken too narrow a view of what they were. His designs were full of faults—the work of a raw beginner by today's standards—recognized as such by their ungainliness. What was disturbing in these designs from the past was not in fact their lack of beauty but their gross design faults which were not recognized at the time. The more 'correct' a design is, the less cause there can be for complaint about its ugliness; furthermore, the creations of the engineer are, little by little beginning to show something approaching beauty. This is beauty, admittedly, of a rather special kind which seems to follow its own rules. It has come about without any conscious effort on our part. In the early days deliberate attempts were made to adorn the face that technology showed the world by borrowing accepted examples of beauty from Nature and Art and using them in an engineering context; this led to levers shaped like swans' necks or machine pedestals embellished with Doric pillars (Fig. 18). Such attempts invariably failed miserably and so the birth of this new kind of beauty is all the more unexpected. Its gradual appearance can be traced particularly clearly from the example of the railway locomotive. Figs 19*a*, *b* and *c* illustrate three stages of development in which the progression towards a particular type of beauty, seen most clearly in the last of these three pictures, is unmistakeable. This special kind of beauty had no place in the world of the past, and neither Nature nor Art have produced anything of its kind; it represents a new element in our concept of what is beautiful, something that we need to study and get to know. Taste for this new kind of beauty has barely begun to develop and one suspects that there is a long way to go.

The engineer has always been intrigued by the question of whether there is any correlation between elegance and fitness for purpose. If we take this to mean that anything appropriate to its purpose will at the same time be beautiful, then there can be no such relationship. Fitness for purpose is limited in time—a 30-year-old design will no longer be considered as answering its purpose, since we now have design resources which differ from those of 30 years ago, and our idea of fitness for that particular purpose will have changed accordingly. On the

Fig. 18.—Gold rolling machine by Krupp, dating from 1840; the ugly shape of the crank arms is typical of designs of this period.

other hand, what appears to us as technical beauty is not liable to change with time; but it is possible to see a clear parallel between the beauty inherent in a design and the extent to which the design approaches technical perfection. Modern aircraft design (Fig. 20) is a prime example of this; even on a small scale, however, we can see that beauty is obviously closely linked with technical sophistication (Fig. 21).

If the appearance of beauty in engineering after the sad experience of the early days came as a great surprise, then it is even more amazing to come across beauty within a technical process, an example of which is seen in Fig. 22. Prompted by these discoveries we may find ourselves wondering whether, in designing something, there might not be a point in consciously trying to achieve beauty—whether, in fact, it may

Fig. 19.—The development of the steam locomotive shows particularly clearly the evolution of a new kind of beauty—technical beauty. (*a*) early, immature design with a comical and grotesque appearance; (*b*) middle stage of development, with the beginnings of a feeling for form; (*c*) later styling, marrying power and elegance.

Fig. 20.—'Concorde'; British Aircraft Corporation.

not be a designer's duty to do so. Knowing as we do how Man reacts to beauty in any form, by a greater sense of well-being and an enhanced efficiency, my belief is that we should do everything in our power to produce graceful designs. But care is called for—trying to 'manufacture' beauty is always full of pitfalls. Ornamentation and other devices for improving the appearance should be used very sparingly; a particularly vivid example of this is provided by the American automobile industry, where a variety of specialists were entrusted with the task of styling car exteriors so as to achieve an elegant appearance. The result was cars which caused a sad shaking of heads all over Europe, and which failed to find buyers precisely because of their appearance. It is really quite understandable that this way of going about things should be a failure; for the styling specialists were not engineers—they were decor designers and commercial artists who went about their work of styling the cars quite oblivious of the intrinsic rules of mechanical design. Beauty in a technical product can only spring from the way it fulfils its purpose, and the way it obeys its own intrinsic rules; and since this can only be appreciated by someone with technical training, it is for him to see to it that the design is a visually attractive one.

Experience so far indicates that the surest way of doing this is to try to get the design technically right—when this is done, beauty follows of its own accord. This can be clearly seen in the gearbox shown in Fig. 23; obviously there has been no attempt at producing a visually attractive design here, since the gearbox will in any case be closed and out of sight when in use. If the truth of this should still be doubted, one has only to think of the example offered by a wrist-watch—no-one can fail to be entranced by the magic of a carefully-made watch movement.

Honest and strict practicality are always a sure guarantee of an attractive appearance in a design, and the same applies to tidiness, a logical utilization of space and a clear layout (Fig. 24). One need hardly mention that elegance in a piece of machinery can play a decisive part in selling the product. It is, of course, true that the technical qualities will be keenly weighed up, and that 1 per cent off the price can sometimes clinch a sale; but it is just as true that attention is still paid to the appearance of the machine.

Fig. 21.—Detail of grinding-wheel support of a Maag gear grinder.

Design and skill in drawing

Drawing is surely the designer's most important working tool. Riedler, in his book on machine drawing[1], begins with the trenchant comment 'All drawing must serve a definite purpose and the way the drawing is done must match that purpose. There is no such thing as drawing without a set purpose in view'. For the designer the purpose of a drawing is twofold. Although it may eventually convey information to others, it is first of all an aid to thought and to arriving at shapes and dimensions, and in this context it is the immediate expression of the creative process as it evolves. As the mind gropes and explores possibilities and forms, the hand adds one feature to another; the lines are vague and tentative, and everything is fluid until its final form is decided on. Drawing instruments are used very little at this stage—occasionally the compasses, a rule even less often. The designer's attention is directed inwards, so that the hand does what is required of it entirely at an incidental level. The drawing that results is of a very special kind, with more of the artist's sketch about it than of the painstakingly precise products taught in technical-drawing classes. What the designer is after is a picture which will reproduce his thoughts as vividly as possible and with a minimum of manual effort. Fig. 25 shows a design sketch by an experienced designer, completed in about one hour. It will be seen that the handling of line is light, and it is very far from being what one understands by a draughtsman's drawing; yet it should be noted that the bevel gears shown are depicted accurately with all the technical detail needed in a design. A draughtsman could, knowing the scale used, take this sketch

1 *Das Maschinenzeichnen*, Springer-Verlag, 1913.

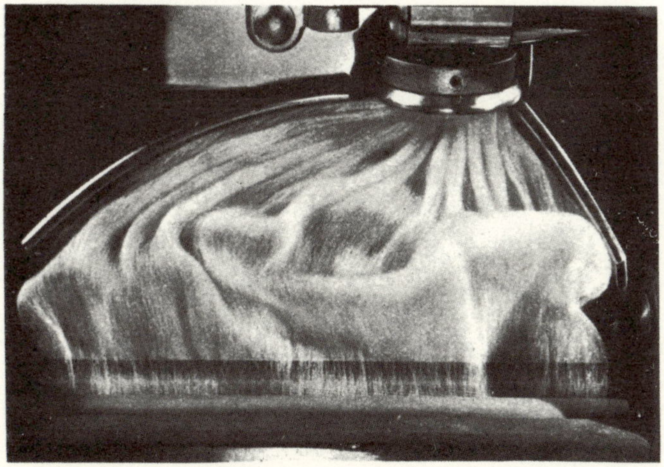

Fig. 22.—Fleece from a combing machine entering the outlet funnel (Rieter, Winterthur).

Fig. 23.—Inside of a gearbox on a packaging machine (Benz and Hilgers, Dusseldorf).

Fig. 24.—Spinning machine, Flyer (Rieter, Winterthur)

as the basis for a complete set of working drawings, and any experienced engineer would be able to gain from it a detailed idea of the designer's intentions.

In contrast with this, the sort of drawing that is taught in technical colleges (Fig. 26) tends to be lacking in expressiveness. Riedler firmly rejects this kind of drawing. 'Mechanically-produced technical school drawings', he says, 'can be found in plenty; all they demand is patience, a little practice and a great deal of time which could be spent to better purpose'.

Developing a special style of drawing is imperative for the designer if he is not to be hampered in his work. We can see how all the great designers of the past have mastered this style; perhaps the most famous example is Leonardo da Vinci. With him it is hard to know which to admire more—his masterly illustration or his capacity for ingenious design (Fig. 27). A great similarity will be seen in the way the drawings in Figs. 25 and 27 have been carried out. It should not be supposed that the artist of Fig. 25 has deliberately taken Leonardo as a model; his style has evolved unconsciously—but since the circumstances surrounding the drawing were the same in both cases, it can be understood how the means of expression show similar features, too. In the case of Leonardo's drawing one is tempted to believe that he made use of a straight edge; but the way the strokes are drawn belies this, and—as can be checked by putting a rule against them—his 'straight' lines are not in fact absolutely straight. Fig. 28 shows another sketch by Leonardo, depicting a gun-barrel seen in cross-section. This demonstrates particularly clearly how close the artist comes to the engineer, for this drawing is from the viewpoint of mechanical engineering, entirely modern and could easily be the work of a present-day designer. Although straight lines of considerable length were called for, no rule has been used. Moreover, the way the end of the tube is closed off is wholly in line with modern knowledge of stress analysis.

Alfred Krupp was another engineer with illustrative skill

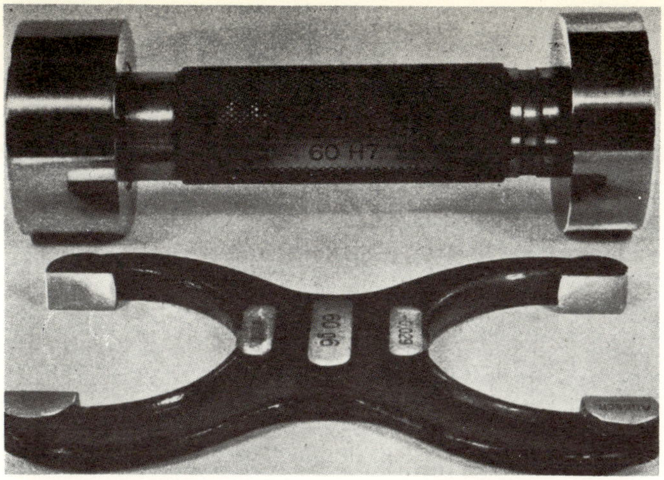

Fig. 51.—Plug gauge and snap gauge for checking sizes of bores and shafts.

buying-out system used by industry, as it has developed today, will only operate properly if the same gauges are in use everywhere. Accordingly agreement was very soon reached in the industrial countries on a system of uniform tolerances, the main features of which were finally brought together by the ISO to form international guidelines.

British Standard 1916 *Limits and Fits for Engineering* covers a range of dimensions from 0·04 to (approximately) 20 in. The corresponding range in the metric system is 1 to 500 mm. Within this range there are, for each measured value, 16 different tolerances which are sometimes known as *qualities* (a term which is not all that clear, since it suggests rather the condition of a surface). B.S.1916 uses the term *fundamental tolerances* and in each size range, numbers 1–4 are intended for precision gauges, 5–11 for machining processes, and those from 12–16 for the remaining shaping processes such as forging, stamping, rolling and casting. The size of these tolerances is governed by a system based on a unit of length—the fundamental *tolerance unit i*—specially created for this purpose. This does not have a constant value, but is a function of the nominal size N in each individual case. Expressed in thousandths of an inch, this is:

$$i = 0{\cdot}001\,N + 0{\cdot}052\,\sqrt[3]{N}$$

with N in inches. Alternatively if i is in thousandths of a millimetre (micrometres or microns):

$$i = 0{\cdot}001\,N + 0{\cdot}45\,\sqrt[3]{N}$$

with N in millimetres. Figs. 52 and 53 show the curves for i against N. By superimposing a cubic and a linear law the tolerance unit for small measured values has been made relatively large. This has been done because a given relative error (variation in the relationship to the measured value) is more difficult to keep to with small workpieces than with large. In order to avoid having to give each individual measured value a definite and individual size of tolerance unit, the whole range of sizes from 0·04 to 20 inches (1 to 500 mm) has been divided into a number of sections for each of which the tolerance unit has a given constant value. This is arrived at through the geometrical mean N of the nominal sizes of the upper and lower limits of the section in question,

$$N = \sqrt{N_u \,.\, N_l}$$

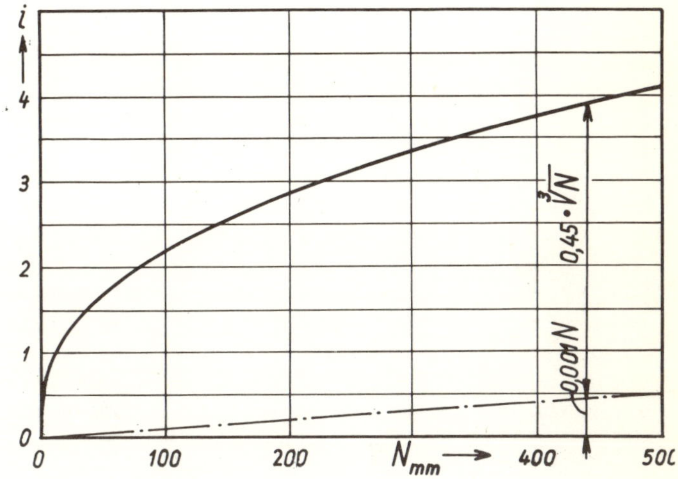

Fig. 52.—Tolerance unit i μm against nominal sizes N mm. Note that the curve is of logarithmic form.

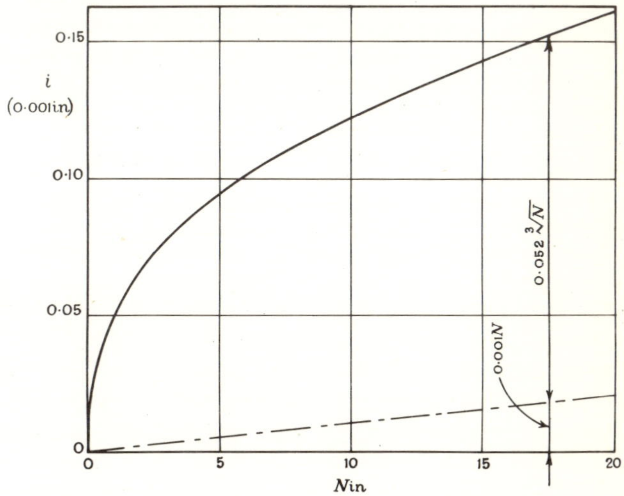

Fig. 53.—Tolerance i against nominal size N in inch units. i in thousandths of an inch and N in inches.

Using this system of length units, tolerances from 1 to 16 were defined, each tolerance being allotted a certain number of length units as follows:

Tolerance	1	2	3	4	5	6	7	8	9	10	11	12	13	14	15	16
No. of units	2	2·5	3·5	5	7	10	16	24	40	64	100	160	250	400	640	1000

It will be seen that we have here a (roughly) geometric series with a step of 1·6 from the 6th quality onwards. See B.S.1916 for tables of the results.

A further and very important problem was how the tolerances arrived at in this way were to be related to the nominal size. At first it seems proper to arrange the tolerances symmetrically about the nominal size, i.e. with equal parts above and below this value; with a tolerance of 0·04 mm (0·002 in.), for example, this would lead to a size being quoted as 50±0·02 mm (1·5±0·001 in.) which would also be reasonable in many cases. In mechanical engineering, however, we need time and time again to combine machine parts in ways dictated by their purpose and circumstances; sometimes there needs to be free play between them, sometimes there must be no play at all and sometimes they must mate with a certain stress between them. The engineer speaks here of *fit*, meaning the mutual relationship between the parts. The fit of a sliding carriage in a guideway, for instance, involves a certain amount of play, so that it can *run*; a shaft is fitted into a hub with virtually no play at all, so that the two parts are *tight* and stationary with relation to each other; the shrunk-on ring of a rotor has what we might term negative allowance, so that there is a pressure or *interference* between the parts. According to the way in which the two components fit together, we therefore speak of a *running fit*, a *tight fit* and an *interference fit*. British Standard 2517 recognizes three classes of fit as follows:

(*a*) *Clearance fit* in which the limits for the mating parts are so disposed that clearance always occurs when any pair made within the prescribed limits is assembled.

(*b*) *Interference fit* in which the limits for the mating parts are so disposed that interference always occurs when any pair made within the prescribed limits is assembled.

(*c*) *Transition fit* in which the limits for the mating parts are so disposed that either clearance or interference may occur when pairs made within the prescribed limits are assembled.

There are basically two ways of arriving at a certain fit between two parts. On the one hand, the tolerance can be arranged symmetrically about the nominal size and the nominal sizes of the two parts varied to obtain the desired fit. If, for example, a shaft needs to have an average play inside its bearing of 0·02 mm (0·001 in.), then the nominal diameter of the shaft can be made smaller than that of the bearing by this amount. On the other hand, it is also possible to give the shaft and the bearing the same nominal diameter, and to move the tolerance for the shaft 0·02 mm (0·001 in.) downwards with relation to the nominal size, and that for the bearing 0·02 mm (0·001 in.) upwards. The question of how such problems of fit should be solved, and of how the tolerances should be placed in relation to the nominal size, called for thorough study and a careful comparison of the advantages and disadvantages of each method. Finally, the second method was decided upon—with the two machine parts having the same nominal size, with a differing distribution of the tolerance with regard to the nominal size. Nowadays most problems of fit are solved in this way.

The fit system based on the international standards recommendations embraces 21 possible relationships between tolerance and nominal size. In the main these are based on simple mathematical rules, which determine the position of the tolerance dependent on the nominal size, and also to some extent dependent on tolerance 7 (see B.S.1916). The various positions are known by the letters of the alphabet, the letters *i*, *l*, *o*, *q* and *w* being omitted, no doubt to avoid confusion. This means that the 16 tolerances chosen can be basically arranged in 21 different ways with respect to the nominal size. This gives a total of 336 possible positions, only some of which are represented in the British Standards; nevertheless the choice offered is an ample one.

It might be expected that such decisions on tolerance position vis-a-vis the nominal size would be definite and once-for-all. It has, however, since been decided to differentiate between external and internal dimensions (i.e. between a shaft and a bore) by on the one hand inverting the sign and on the other by also deriving the position of the bore tolerance from the tolerance of the shaft of the next lower quality. This means that different symbols have to be used for internal and external dimensions. Capital letters were adopted for indicating internal dimensions, and lower-case letters for the external dimensions. The indications for a shaft, for instance, will run '50—*n* 6' and for a hub '50—*N* 6'; this means that 50 is the nominal dimension, *n* and *N* are the positions of the tolerance, and 6 its size. The approximate position of the tolerances with respect to the nominal dimension can be seen from Fig. 54. Special letters in the series are *h* and *H*, which always lie exactly on the nominal dimension from the material side.

Carefully thought out though this system of fits is, it has its faults. In recent years there have been repeated proposals for a modification, referred to by the term *mean tolerances*. This move has been prompted by the realization that the actual dimensions of workpieces manufactured to tolerances in most cases lie closer to the GO side than to the NO GO side. Fig. 55*a* shows the approximate frequency distribution within

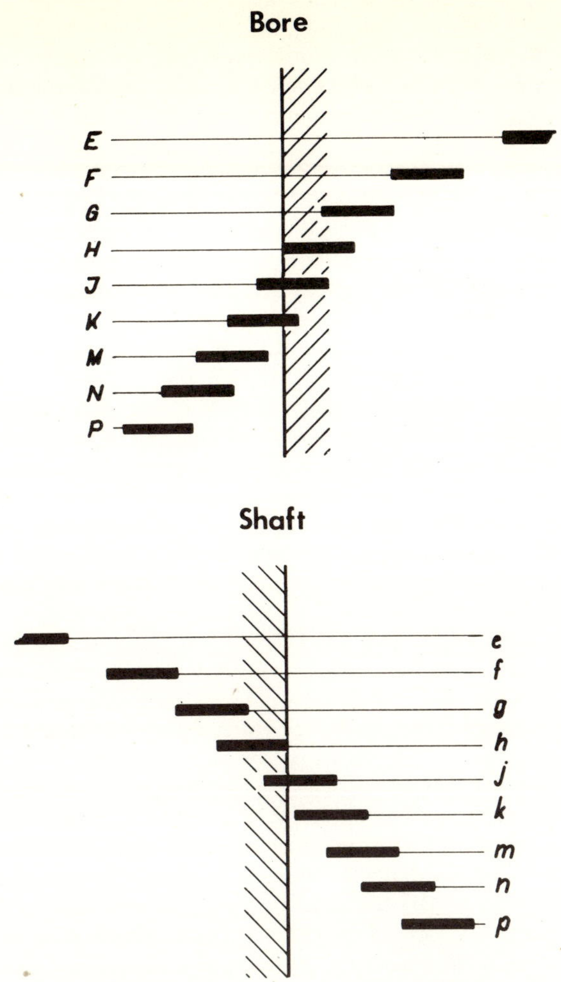

Fig. 54.—Relative position of the tolerance zones close to the nominal dimension.

the tolerance range for internal and external dimensions. This tendency comes about because the machine operator stops machining as soon as the dimension begins to exceed the GO side of the gauge, so that the actual dimension is usually very close to the GO dimension. His 'terms of reference' give him no reason to take the machining of the item any further than this. To overcome the disadvantages of this wholly understandable approach (which is dictated by caution) it has been suggested that instead of the GO gauge we should use the NO GO of the following tolerance zone—for example, to use position *N* instead of *M*, as a result of which the point would be brought back from the GO side to the centre of the tolerance zone. It is argued, however, that this will simply result in work to the next higher quality, since it basically—like this higher quality—restricts the possible range for the actual dimension, and forces the operator to get closer to the 'dangerous' NO GO side. The shortcoming could also be overcome if the designer, in setting his tolerances, were to make prior allowance for the fact that the majority of the workpiece lies on the GO side and were to select the position of the tolerance zone so that the amount of play he is aiming at comes in the middle in spite of the shift we have described (Fig. 55*b*). As can be seen, all that is needed is a slight separation of the tolerance zones. A disadvantage that would have to be accepted with this procedure would be an occasional case of the play being too great when two workpieces coincide both of which lie on the NO GO side of the tolerance.

The reason for this shortcoming lies deeper however; it is tied up with the principle of the fit system, which is based on

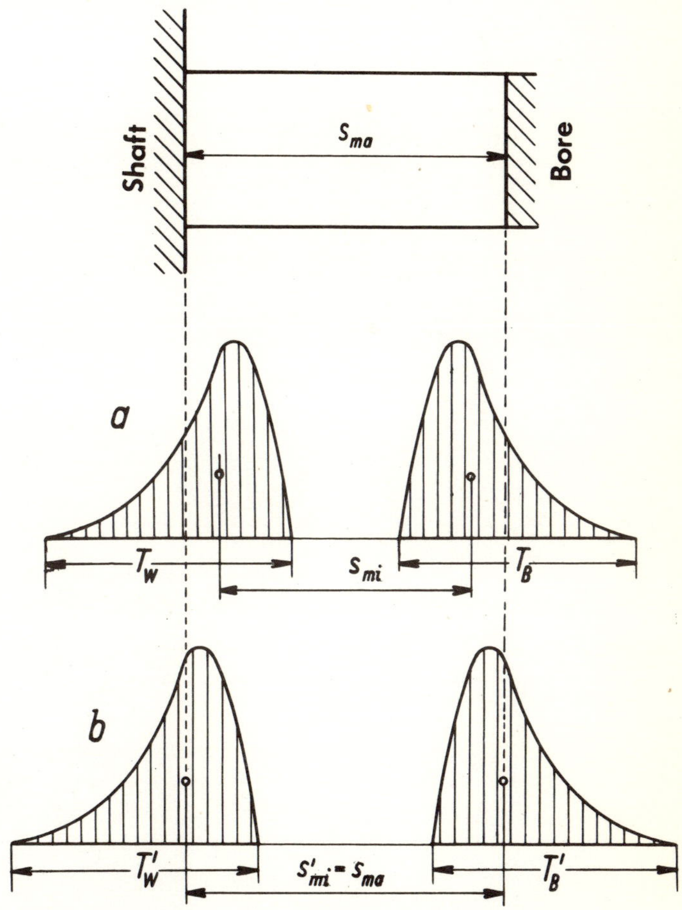

Fig. 55.—(*a*) Frequency distribution of actual size in the tolerance zone; (*b*) shift of tolerance zone for matching to the desired average play.

matching two parts to each other by aiming for one given size (the nominal dimension) which really has nothing at all to do with the fit. If, for instance, we want a piston to run in a cylinder with 0·01 mm (0·0004 in.) play, then it is not basically a matter of whether the diameter to which the two surfaces are to be related is 100·0, 100·2 or 99·7 mm (4·00, 4·01 or 3·98 in.); in a case like this it would be quite correct, and the simplest way of going about things, if the bore were first made at approximately 100 mm (4·00 in.) and the piston then matched to this. The care to be expended on fitting the two together would then only have to apply to one of the two parts. The trouble is that this would do away with interchangeability, and so much importance is attached to this factor by modern industry that the burden of two-fold fitting work is accepted.

At all events, the example of a piston and cylinder we have just quoted provides a warning not to exaggerate the variations in the tolerance zone when manufacturing to given fits. It is possible to obtain a required fit by leaving the position of one of the two tolerance zones alone, and making the variations only in the other. Depending whether this is done on the hole or the shaft, the engineer speaks of a *unit-bore* system or a *unit-shaft* system. Most firms work to one or other of these systems, and advantages are to be found in each depending on the individual circumstances of the job. If we are dealing with a longish shaft which is to run in several bearings and carry a number of gearwheels, then it will be better to give it the same diameter throughout its length and to match the bearings to it by varying their tolerance zones, rather than the other way about. On the other hand a sliding control valve which passes through a casing with a varying amount of play at the various inner walls will be better made with differing diameters, and the casing with the same internal diameter throughout.

Here, again, it is always the designer's job to weigh up the pros and cons carefully so as to find the most suitable solution. And so we can see that even the existence of all the paraphernalia of a tolerance system and a system of fits still does not relieve him of the task of looking at things, and thinking about things, for himself.

Limitations and changes in the production process

The profitability of mechanical engineering depends to a very large extent on whether, during manufacture, a design can follow the normal path through the production departments. Anything that is out of the ordinary, for any reason, entails extra costs; in fact it starts to do this as soon as it requires special attention from those running the business. One should, therefore, try to design in such a way that ordinary materials (such as cast iron, sheet steel and steel sections) can be used, and be dealt with by ordinary machine-tools (such as milling and grinding machines, lathes, drills and the like). The designer must know what machinery his firm possesses, and what its limitations are, and he will be well advised to call on the advice of the appropriate manager where matters of workshop production are involved.

Most of the firms making machinery earn their profits in the broad middle range of ordinary everyday production work which calls for neither special machinery nor specially-trained operators. And it will be all the easier to get sub-contracted work accepted if it fits into this 'average' category. On the other hand it is becoming increasingly common for particularly stringent requirements to be laid down for a given product, in which case the customer is quite prepared to pay a correspondingly high price. We might, for example, mention the whole field of rocket engineering and space technology for military and civilian purposes, where the question of cost comes right at the end of the various considerations. In such instances the success of a design often depends to a large

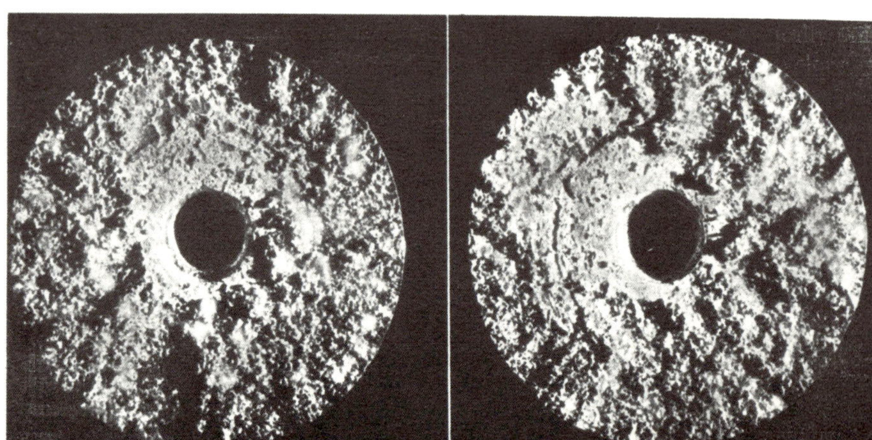

Fig. 56.—Rod hard-soldered by the Hyde process, after fracture test. The fracture surface runs for the most part through the original material, alongside the joint.

extent on whether one particular item or another can, in fact, be made at all; then the question of how it is to be made hardly matters any more. With an eye to circumstances like this, the designer must make a thorough study of exactly what his firm is capable of manufacturing, and at what point the limit of its capabilities is being exceeded. Today there are a whole range of such borderline areas in manufacturing methods, some of which we shall now discuss.

One notable example is Hyde's hard-soldering or brazing process. Soldering is the union of two workpieces by means of a bonding metal (the 'solder'). In hard-soldering, which gives a specially strong joint, copper is used as the bonding metal (known as 'spelter'); the strength obtained in the joint is usually equivalent to that of the basic metal involved—indeed it is often greater—for fracture tests have frequently shown that the break has occurred in the workpiece itself, alongside the brazed joint (Fig. 56). The urge to use a brazed joint is therefore very great; but it is matched by the difficulty of producing a clean joint by this method. The spelter will only fuse with metal which is free from oxidation, and with copper's high melting-point of 1100°C this has meant until recently using a large amount of flux. Consequently, ease of access to the joint and the possibility of subsequent corrective work are essential for good brazing. This means that there are strict limits on the use of the process; and these seem to apply just where the use of the method is most attractive. Take, for example, the sliding control valve seen in Fig. 57—how are we to cast this, when it is barely 100 mm (4 in.) long? Turning it up from a solid piece would be just as hopeless, while fabricating it from separate parts welded or hard-soldered together would be impossible on account of the inaccessibility of the joints. This control valve is, nevertheless, typical of a whole series of similar jobs; obviously, there was a gap which needed to be filled. This was done when Hyde carried out the brazing in a protective atmosphere of gas which prevents the formation of oxides on the workpiece at high temperatures (and which even removes any oxidation already present). The gas used for this protective envelope is hydrogen which, because of its affinity for oxygen, immediately reduces any metallic oxide formed at high temperatures. Since we are able to employ electrical radiant heat, there is no need to have a naked flame at the point of fusion. The process, after a long period of development, is now as follows: the parts to be brazed are fitted together with only a slight amount of play, and the spelter either pushed into the joints in the form of a foil or attached to the workpiece itself, close to the joints, in the form of wire (see the arrows in Fig. 58). The workpiece is then placed in a heat-proof and gas-tight container (usually using special clamps), which after thorough evacuation of the air is filled with hydrogen. Finally the entire contents are heated electrically to 1100°C, at which temperature the spelter melts and is drawn into the joints by capillary action. The workpiece is then allowed to cool, and the copper fuses with the surfaces at all the joints. Apart from a very slight heat distortion, the size and form of the workpiece are unaltered, so that quite small tolerances on the fitting surfaces are enough for final machining. The surface of the workpiece will be sparkling clean and occasionally, if an excess of copper has been used, there will be an extremely thin coating of copper which forms excellent rust protection. There is a vast scope for this very convenient method, which makes it possible to assemble highly complex components from simply shaped parts and to fuse them together to form a whole by a quick process which demands no manual skill (Fig. 59). The process was found to offer special advantages for the fabricating of rotors like that shown in Fig. 60, where the piston for absorbing side thrust always had to be made separately so as to leave the

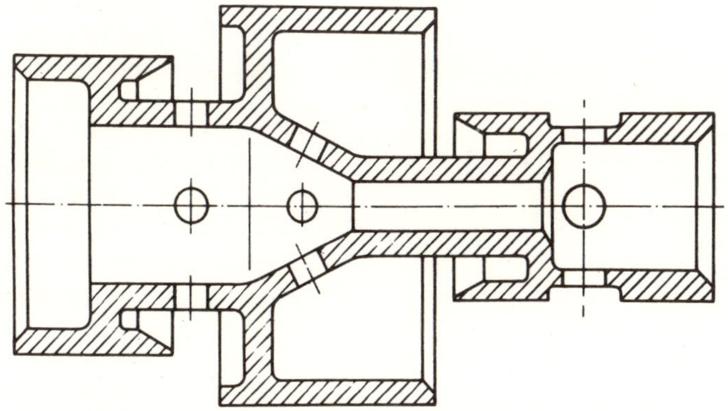

Fig. 57.—Sliding control valve (approximately 100 mm long); hardly any of the normal manufacturing processes are suitable for producing this component.

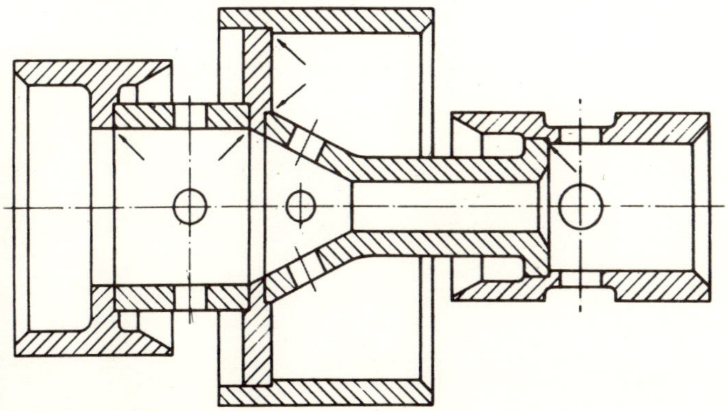

Fig. 58.—By the Hyde process, the work is built up from separate pieces of plate and tube and hard-soldered at the appropriate points (arrowed).

underside of the rotor free for riveting on the vanes. By hard-soldering the vanes instead of riveting them, it becomes possible to produce the base of the rotor and the piston in one piece (right-hand side of Fig. 60).

The plant needed for this process is not, however, simple; it is essential to prevent any contamination of the hydrogen with the surrounding air, as this would result in an explosive mixture, and the method calls for great care and a thorough checking at each stage in the process. Fig. 61 shows a plant of this kind.

The best, and most sophisticated, way of producing a piece of machinery is still to make it by hand. The human hand can find a way of doing something when all mechanical efforts have failed. There is no planing machine, no grinder and no miller capable of producing a flat surface with as many contact points as can be made by hand with a scraper, and consequently things like high-precision crosshead guides and bearing surfaces can only be produced with a sufficient degree of accuracy if they are handmade. The amount of unevenness on the surface to be scraped is measured with a surface plate or cylinder (Fig. 62) and any areas which stand proud are shaved down one after another with a scraper (Figs. 63 and 64). The degree of precision reached nowadays is within 0·001 mm (0·00004 in.). Working to this degree of accuracy requires comprehensive measures in order to create suitable conditions—the workshop must, for instance, have a system of air-conditioning which will ensure that the variation in temperature is no greater than 1–2°C in the course of a day.

What distinguishes working by hand from any kind of machining is the virtually unlimited possibilities it offers for producing various shapes. We might say that anything and everything can be made by hand, and a comparison with the sculptor (who as we know creates the purest forms) is inescapable.

Fig. 59.—Valve housing and differential piston, hard-soldered by the Hyde process. They are cut open to show how the joints are completely filled with spelter.

Fig. 60.—Rotor from a radial compressor. Left: base of rotor and piston made up separately. Right: base and piston made as one unit. This has been achieved by hard-soldering the vanes to the base instead of riveting them.

Fig. 61.—Plant for hard-soldering by the Hyde process. The furnaces are let into the floor of the workshop. The one in the foreground and another at the back are in use, and are showing a hydrogen flame. A constant excess of H_2 gas, which burns to give a visual indication, is provided to prevent any air being drawn in from outside (*Photo: Brown Boveri, Baden*).

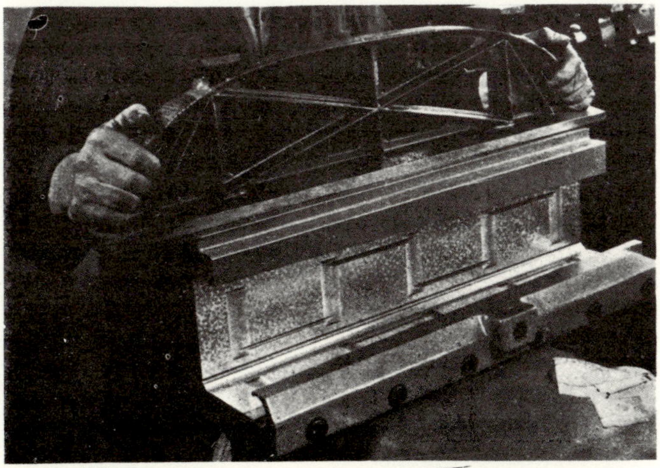

Fig. 62.—Surfaceplate for checking the flatness of a surface on the machine-tool which is to be scraped by hand. To prevent any buckling of the plate, it is stiffened by a truss (Messrs. Dixi, Le Locle).

There is so far no mechanical device or machinery capable of finishing Pelton turbine blades. These are first produced as castings, with a fair degree of accuracy, and then worked to the required shape with a hand grinder (Fig. 65). The operator is told this shape by means of a number of cross-sections marked

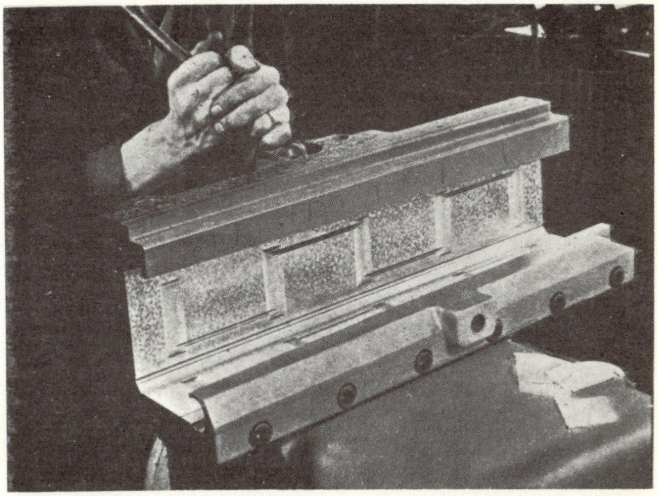

Fig. 63.—Wherever highspots are shown by the surfaceplate, the metal is removed by scraping over small areas. For the surface to be acceptable, there must be 40 contact points per square centimetre.

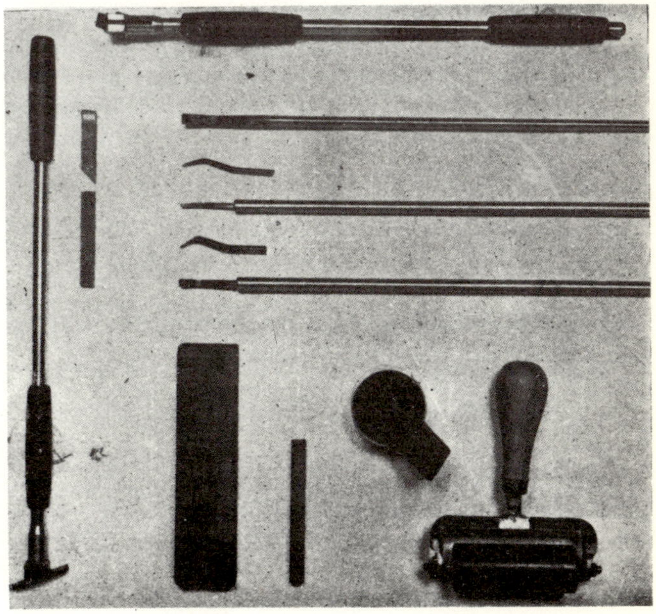

Fig. 64.—Complete set of scraping tools (Dixi, Le Locle).

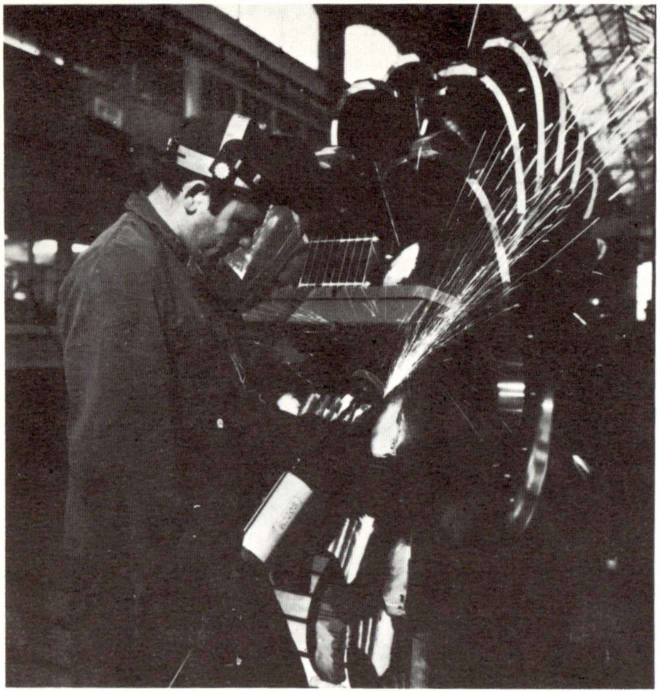

Fig. 65.—Machining the inner surface of Pelton turbine blades with an emery wheel. The tool is guided solely by the skill of the operator. (*Photo: Escher-Wyss, Zurich.*)

on the workshop drawing, and he makes repeated checks with a template (Fig. 66) to see how close he is to the final shape. This shaping is made far more difficult when the rotor has been made as a single casting, when the inner surfaces of the blades become less easily accessible (Fig. 67).

One of the most difficult jobs of all is making banks of nozzles for a steam turbine, working from the solid. For low and medium pressures, these components are made up from two parts, each of which is machine-turned or ground to the required shape. With a very-high-pressure turbine, however, a nozzle of this kind would not stand up to the loads put on it, and consequently it must be made from a single piece of metal. Depressions are first made in the steel block by drilling, and these are then widened, chiselled, filed and polished to form nozzle channels passing right through (Fig. 68). Since high-grade material is used for such nozzle banks, only the best tools are adequate for this exacting work, and the cost of producing the nozzles is so high that they end up almost literally worth their weight in gold.

Another amazing example of how hand-working can provide the only possible answer to a problem is seen in the Hirth design for an assembled crankshaft (Fig. 69). One of the greatest successes in aero-engine construction came through the use of enclosed bearings at the big-end of the connecting rod.

The problem is to find some way of fitting a bearing of this type over the crankpins of the crankshaft; the answer has been to split the crankshaft itself instead of the bearing, bolting it together again once the bearing has been fitted over the pin. The bending loads at the point of junction could be coped with by a simple superimposing of compressive and tensile stresses, and would allow of smooth mating surfaces; but the torsional loads would call for an interlocking surface where the two sections of the shaft join, i.e. for some kind of serra-

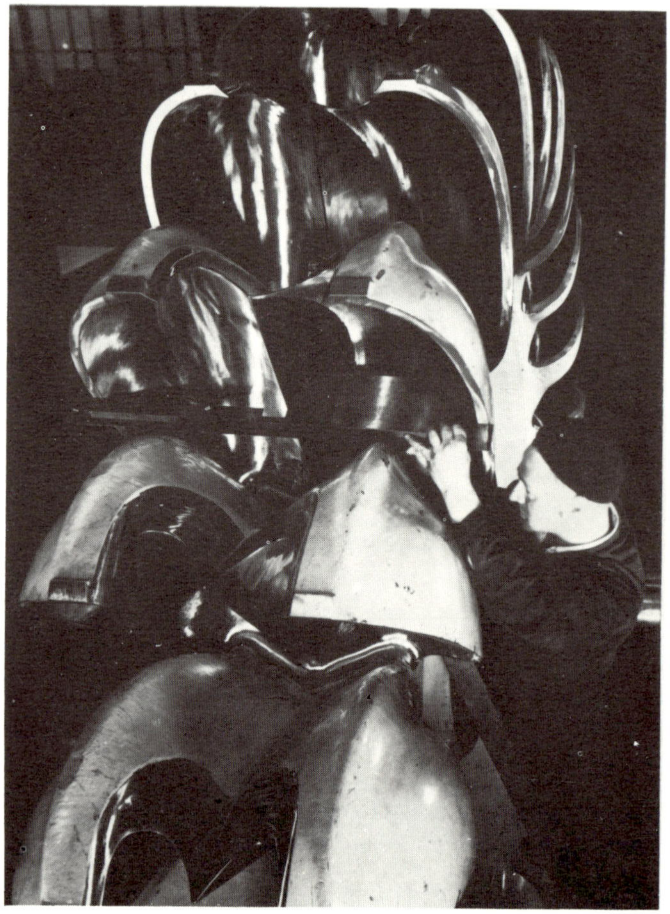

Fig. 66.—Testing the reflection angle (Escher-Wyss, Zurich).

Fig. 67.—Complete turbine rotor as a single casting. Difficult to machine, as access to the inside surface of the blades is restricted. 60 MW, head 320 m, diameter 3750 mm (Escher-Wyss, Zurich).

tion. This is the crux of the problem, since there is no machine which will cut such a series of serrations so accurately that there is absolutely uniform loading on the side faces of the teeth. This means that they have to be matched up by hand—a job calling for extremely painstaking skill, since the tiny teeth have to be made so that there is complete contact all over their surface.

With a large production run, hand working becomes too costly; it is, however, worth building quite large machines which are to a large extent matched to the special features of the item being produced; this still allows us to manufacture economically even very complex forms. Copying methods have made great progress in this field; here, a model or 'pattern' is prepared of the component to be manufactured (this is usually done by hand), and a feeler then follows the contours of the pattern and guides the cutting tool. The low-pressure vanes for a steam turbine are, for instance, ground in this way (Figs. 70 and 71).

Such manufacturing processes are inevitably expensive; at the present time they are still in their infancy, and it is impossible to predict what the final stage of their development will be. Electronics has now come into the picture, with a computer being used to analyse the data for the surface to be

Fig. 68.—Nozzle segments for high-pressure steam turbine, in stainless steel, made from the solid. Although these are extremely difficult to produce, there is no irregularity to be seen (Brown Boveri, Baden).

Fig. 69.—Aero-engine crankshaft which can be dismantled; Hirth serrations are provided at the mating surface and on the set bolts (cf. bolt in Fig. 33).

reproduced, and these figures then being used to control the tool. It is particularly valuable, for example, in an aerofoil wing, to determine the mathematical laws for the mean camber line and that of the profile separately, and leave it to the computer to superimpose the two. It is also quite conceivable that the grinding of gears will eventually be carried out in the same way, which would obviate the need for using an involute tooth form.

We need hardly stress that engineering production plant and processes are in a state of constant change—over the last 20 years several hundred new manufacturing processes and possibilities have been discovered. They include one specially noteworthy range of related processes which can only be described as a revolution in production technique. During machining with metal-removing tools—a process to which practically all machine parts with any pretension to accurate dimensions have had to be subjected—we have had (rather like peeling an apple) to cut away the outside layer which in most cases is the best of the metal. In castings, drop-forgings and rolled sections, this outer layer invariably has greater strength than the metal inside the workpiece. For many years this state of affairs went almost unnoticed; at all events, it was hardly ever discussed, and only a few people were aware that any application of a cutting tool to a cast, rolled, or stamped workpiece immediately cuts down its load capacity by a substantial amount. A swivel axle for a motor-car which has been machined from the solid is beyond any doubt far weaker than one that has been stamped or drop-forged. And so we have the strange position where it is in fact the most costly and time-consuming manufacturing process—machining—which detracts from the strength of the finished product. Methods which use the plastic properties of the metal to shape it (e.g. casting) give the workpiece a surface structure which is extremely well-suited to withstanding mechanical loads, and which moreover offers greater resistance to corrosion. A doubly important aspect of this is that these two factors (mechanical stresses and corrosion) always have their greatest effect at the surface—we have only to look at the typical distribution of torsional and bending stresses to see this.

The fact that cavities, graphite inclusions and other flaws occasionally occur in the workpiece during casting, forging and stamping cannot be looked upon as an out-and-out disadvantage of these processes, for these flaws are invariably found inside the workpiece where (as we have just pointed out) the mechanical loads are not all that high and where there is no wear or corrosion. Destruction of the material, and fractures, always start at the surface and never inside the metal (unless a flaw happens to coincide with an area carrying a higher stress), and consequently we have everything to gain by leaving the surface layers undisturbed. Where we have broken them down, this has obviously not been done through sheer pleasure in destruction, but because of the insistent need for keeping to prescribed dimensions. We plane, grind or mill things like gears only because casting or stamping will not give us the desired degree of accuracy, and we turn and grind axles and shafts because we cannot get their diameter accurate enough by rolling or drawing the round steel section.

Some time ago, however, it began to dawn on people that by machining away metal we were losing something very valuable, and we now try to give a workpiece its final form during a work process that will provide this valuable and desirable surface structure. In the past, steel tube has been made with a dimensional accuracy of a few millimetres or tenths of millimetres, and subsequent machining has been unavoidable; now attempts are being made to achieve greater accuracy in production. Firms such as Accles and Pollock or Phoenix-Rheinrohr have been working on this problem for some ten or 20 years, and their methods of cold-drawing metal are today so sophisticated that they can supply steel tubing with a dimensional accuracy of 0·05 mm (0·002 in.). Accuracy of this order is fully comparable with that obtained by turning or grinding a cylinder, so that it is possible to incorporate sections of this tube in a machine without any further processing. Tubing of this kind is, for example, being used for cylinder jackets or for plunger pistons in hydraulic presses (Fig. 72); that this is possible will be appreciated when we learn that the average surface roughness value is about 0·001 mm (0·00004 in.). This excellent figure does not come from any kind of machining, but from additional cold-forming on a lapped steel arbor. The amount of care that goes into the manufacture of these tubes can be judged from the fact that a constant check is kept on the output with surface-testing equipment (Fig. 73).

Fig. 70.—Automatic copying machine for milling three LP steam turbine blades at a time, with hydraulic control (Starrag Rorschacherberg).

Fig. 71.—Finished turbine blade for the copying miller shown in Fig. 70 (polished after milling). A high-precision product, which is as much a work of art as it is a piece of hand-produced machinery.

For quite a long time threads which are likely to be subjected to extreme loads have been rolled rather than cut or turned (Fig. 74). Depending on the kind of metal used, the improvement in quality may be several times that of a cut or turned thread. Alongside these processes, which are continually being improved, there is growing up a whole technique of precision stamping and extruding workpieces which are then ready for use without further machining; these are known by the generic term 're-forming techniques' (O. Kienzle). In the past lubricating grooves in shafts or sleeves were milled or cut by hand; now they are beginning to be stamped into the workpiece. Fig. 75 shows a sleeve with stamped helical grooves, designed to take a similarly-shaped shaft; these grooves have been produced by, as it were, squeezing the sleeve over an arbor in the cold state (Fig. 76). A great many small to medium-sized components are now made by a similar method known as *extrusion moulding*.

When we look over the whole field of engineering we can see that the dividing lines which used to separate one sector from another are beginning to become blurred, and that there are more and more transitional stages between hammer forging and drop forging, hot pressing, extrusion moulding, cold forming and stamping.

Fig. 73.—Testbench for inspecting surface finish with the Perth-O-Meter.

Fig. 72.—Hydraulic presses with cylinder and plunger piston made from precision-rolled steel tube (*Photo: Phoenix Rheinrohr AG*).

Fig. 74.—Thread rolling machine. The roller head is on the right to the front, and the axis is adjustable to suit the thread angle (E. Grob, Männedorf).

Probably the most amazing example of the cold-forming technique is to attempt to roll gearwheels and spline shafts. We can only describe this as a risky venture when we think of the accuracy that is needed at the bearing surface of the gear teeth to obtain quiet running and satisfactory meshing; the extremely long life of gears made this way does, however, seem to make the trouble worth while. Looking back over engineering history, we find that the earliest gearwheels were made from wood, and were later replaced by castings; castings were not, however, adopted with any thought of improving the structure of the working surfaces, but quite simply because this was the cheapest way of making gears. The precision left much to be desired; they were noisy and quite unsuitable for providing an efficient drive by modern standards. Today things have changed. If we consider using non-cutting shaping for gearwheels, it is solely with an eye to the improvement in

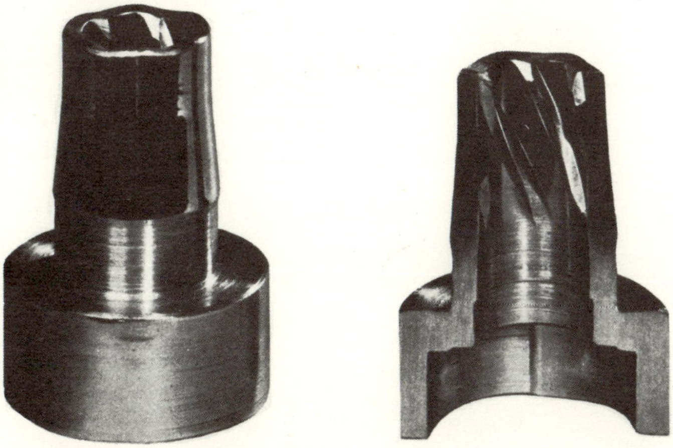

Fig. 75.—Sleeves with internal helical grooves produced by pressing over an arbor when cold (Cincinatti Milling & Grinding Machine Inc., U.S.A.).

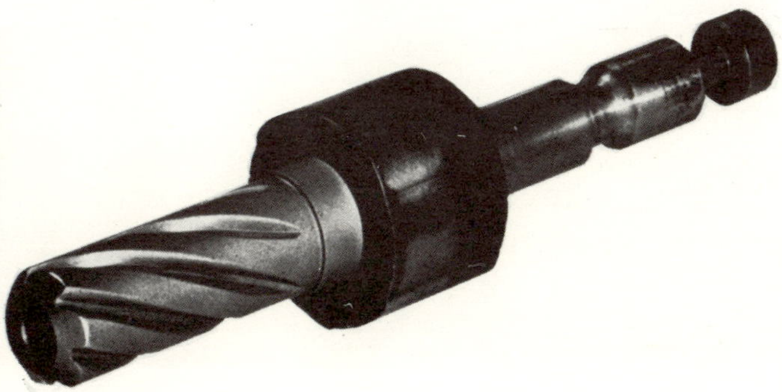

Fig. 76.—Arbor for producing the internal surface of the sleeves seen in Fig. 75.

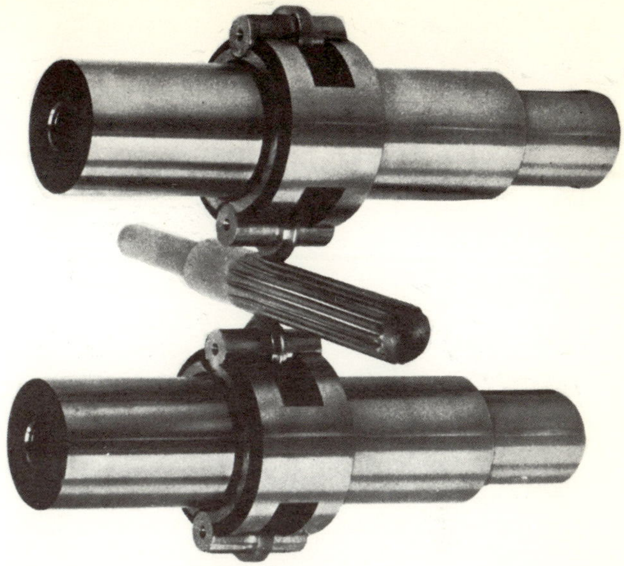

Fig. 77.—Roller carriage for cold-rolling spline shafts, etc., showing the principle of the machines by E. and B. Grob.

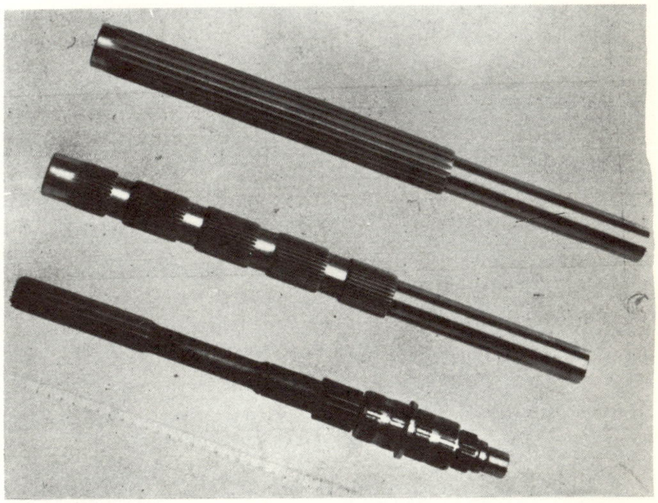

Fig. 78.—Typical examples of cold-rolled parts made by the E. and B. Grob process.

structure in the highly-stressed peripheral areas—down the sides of the teeth and at their base. Two brothers, E. and B. Grob, were the first to use the rolling method for making gearwheels. Steel rod was passed through rollers to make a spline shaft, and this was then chopped up into separate discs. Nowadays we work with a rotating head carrying two diametrically-opposed rollers, which work alternately to shape

Fig. 79.—Gearwheel produced by cold rolling.

the blank a little at a time as they move along the shaft. Fig. 77 shows the principle of this machine, and Fig. 78 illustrates some of the items it can produce. Gearwheels made by this process are very accurate (Fig. 79); dimensional errors on the sides of the teeth are within 0·01 mm (0·0004 in.) and the roughness is in the region of 0·0003 mm (0·00001 in.). The edge life of the rollers in this type of machine is quoted by the length of section produced, and at the moment stands at around 1000 metres (3000 ft).

These machines are the very cream of machine-tool design, and are understandably very expensive (Figs. 80 and 81). They work very fast—an average length spline shaft is produced in about 10 seconds—so they do pay for themselves, especially on jobs where the metal is to be stressed to the limit (e.g. in gearbox parts for motor-cars, or equipment for the armed forces). Fig. 82 shows the excellent structure of the highly stressed area round the foot of a gear-tooth. Metals with an extreme yield point are suitable for this method, and steels with a breaking stress up to 10^9 N/m^2 (150,000 lbf/in^2) can be satisfactorily handled.

In the foundry, too, new techniques for precision casting are being evolved, with sand increasingly replaced by iron

Fig. 80.—Overall view of the E. and B. Grob cold rolling machine.

Fig. 81.—Close-up of the rolling section of the machine seen in Fig. 80. The roller heads are in the centre section, and the finished spline shaft can be seen to the right.

chills which are preheated to almost the melting-point of the metal being cast before this is poured into the mould. Careful prior calculations of the amount of expansion in the mould and in the casting itself make it possible to achieve a dimensional accuracy which allows the parts to be used with hardly any finishing work. To prevent cavities and gas inclusions the molten metal is injected into the mould under pressure (whence the term *injection moulding*), or centrifugal force is used by spinning the mould (i.e. centrifugal casting). Pressures up to 100 atmospheres are not unusual, and Fig. 83 shows a machine part made this way.

Anything requiring a medium degree of precision can be cast in chill, and this covers a vast range of machine parts such as levers, rollers, and wheels, and components for couplings, bearings and valves.

A particular advantage found in this casting process is that it is no longer necessary to grind the seating surfaces for bolts, since the precision obtained is already sufficient to give a firm contact area for bolt heads and nuts. We shall show later how absolutely essential it is, when casting with sand, to machine accurately all the surfaces which come into contact with other parts, including all the seatings for bolt heads and nuts.

All these modern techniques have, in the course of their development made one thing quite plain: every way of circumventing the machining of the workpiece with a cutting

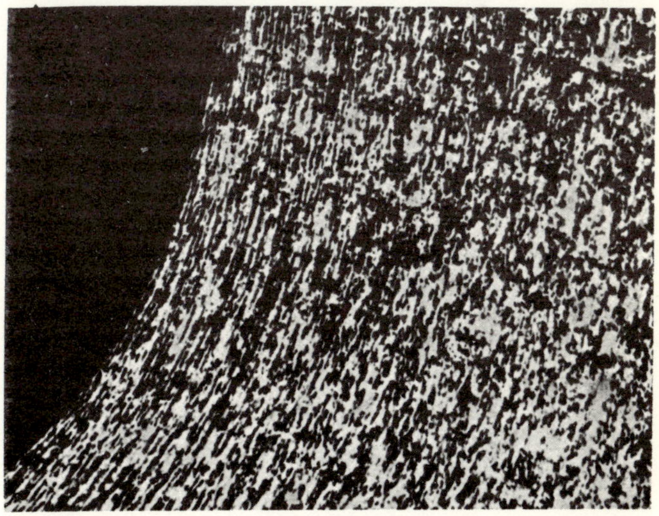

Fig. 82.—Metal structure at the foot of a geartooth cold-rolled by the Grob process. The gneiss-like stratification of the metal is increasingly evident as the surface is approached.

tool entails a fair amount of cost in equipment and plant of various kinds. Fig. 84 illustrates a press used for forming large workpieces, and the man in the picture shows its immense size. We might also mention here that it is sometimes necessary with extrusion presses to pack the extremely hot metal in glass wool, to provide a measure of lubrication and to prevent it from damaging the dies at these extreme temperatures.

So we can understand why, alongside these developments, efforts are constantly being made to find some way of making good the damage done to the workpiece when the surface is removed by skimming, grinding and the like. This is to some extent possible by subsequent finishing of the turned or ground surface with hardened and highly-polished steel rollers. This has brought a substantial improvement to the strength of cut screw-threads, and cases have been known where it has raised the prolonged alternating-stress strength of the thread groove from 7×10^7 N/m^2 (10,000 lbf/in^2) to 17×10^7 N/m^2 (25,000 lbf/in^2).

Similar success has been achieved with the furnishing of grooves in a shaft chamfer and the like, and with the drifting (driving in of polished and hardened steel drifts) of the oil passages of highly-stressed workpieces.

Fig. 83.—Cam disc made by the pressure-casting process for a machine-tool in the Daimler-Benz works, cast by Zentroguss GmbH, Hirzenhain, Upper Hesse.

Fig. 84.—Mammoth press for cold-forming heavy-section sheet steel (Siempelkamp, Krefeld).

The most amazing success of this kind has, however, been achieved by using negative-rake angles in turning and milling, that is to say making the cut so that the metal being removed is not lifted off the workpiece but pressed against it. Angles of 40°–50° have been used. The depth of cut is very small (in the region of 0·5–0·7 mm (0·02 in) and the tool feed rate is also low. In this way residual stresses of around 10^8 N/m^2 (15,000 lbf/in^2) are produced at the surface of the workpiece. The process is still a recent one, and has not been developed to any great extent; but the results achieved so far are very promising.

Control and self-control

The building of a crane ends with a loading test, and that of a machine when it is run-up on trial. There is no doubt at all that these are the decisive steps separating production and putting-into-service; but testing and inspection stretch well back into the manufacturing process itself. Rotors which will run at high speeds are given their first mechanical test as soon as they come off the lathe, when they are run for a fairly long period at excess speed. Depending on the average running speeds likely in operation, centrifugal accelerations up to 9 *g* may be involved, and Fig. 85 (which shows the results of the disintegration of the rotor of a 45,000-kVA generator) is proof that, though these tests may be carried out as a routine measure in most factories, they are not done just as a formality. Later, when the first test has been passed, come electrical, heat and hydraulic tests. Likewise, coolers and pressure vessels are tested for strength and gas—or water—tightness before being installed.

It is not by accident that testing occupies so large a place in engineering practice. The chances of doing something right are limited, while the chances of doing it wrong are almost endless. To end up with a satisfactory product within a reasonable period of time therefore calls for the consistent co-ordination of all the people and plant concerned, working towards one and the same objective, and is only possible through a carefully-planned system of control. In the first instance, such a system entails a check on the work carried out, to see that it is being done correctly and matches the instructions given by the workshop drawing. It also, however, involves a searching inspection of the materials and a final inspection to see that there are no undetected flaws in the material.

It would be a mistake to think that testing and control apply only to the workshop and the material used. An experienced designer will admit that he himself is the most likely in all the long list of possible sources of error, and he must first of all be suspicious of himself in everything he does. Not long ago the Case Institute of Cleveland, Ohio, financed by the Foundry Research Foundation, carried out tests on ten different and common types of cast machine parts. The most important discovery was that the main cause of the fractures that had occurred was not flaws in the casting, but design faults. Admittedly 90 per cent of all the castings had casting defects, and yet they would withstand from 4 to 20 times their operating load. When fractures occurred, these were in the great majority of cases at points where there was a concentration of stress due to poor design. Avoiding such concentrations of stress is far less a problem of materials than one of design. Another noteworthy observation in the Case Institute investigation was that the fractures occurred even though the safety factor had, with most of the parts concerned, been set extremely high.

Fig. 85.—Test-pit after the explosion of a turbine rotor at 3700 rev/min, due to a flaw in the material inside the rotor. The extent of the damage will give an idea of what the consequences of an explosion like this could be if it occurred in an open room.

Self-control, then, is for the designer the most urgent need in this whole system of checks, and the sooner it is achieved the more effective it will be; any following of a wrong path will then be stopped as soon as the first false step has been made. There are designers who can manage to bring a design to fruition working on their own; they are to be admired, but they are unfortunately few and far between. Most designs are the result of teamwork, and one happy circumstance in this is that the person leading the team always has to explain his ideas to his collaborators before they can be put into effect. This constitutes an excellent safeguard against faulty thinking or weaknesses in concept, since explaining something clearly

Fig. 86.—Railway engine axle prepared for a test run. The measuring leads run to the measuring apparatus inside the control van.

to others demands a clear formulation of one's own ideas and a careful thinking-out of the details.

Control must, however, continue to be exercised while the design is being worked out, for often it is only when the detailed planning is being done that many of the problems come to light; often, too, they come to light by chance. Of course things should not be left to chance—a careful systematic search must be made for every possible error, and many designers have evolved their own system for doing this. Many of them use a checklist to make sure that they have not missed anything, and as the design is developed they cross off one item on the list after another. One disadvantage of a system of this kind, however, is that it obliges the designer to cast his net in an area where he already knows that there are no fish to be caught.

The most important of all the stages of inspection is the acceptance test, and the test running that precedes it. Everything of importance is checked; the machinery is often covered with a maze of test leads, and a whole team of observers are at work keeping a constant watch on instruments such as pressure gauges, thermometers, columns of mercury, voltmeters, wattmeters and so on (Figs. 86 and 87). The number of readings to be taken may be so great that it is easier to photograph all the gauges and dials at one go and analyse the results from the photographic prints.

This analysis may take days. The results are rushed to the design offices, and it can make the designer's day when the figures are all as they should be; or, of course, spoil it if they are not! If the machinery does not give its guaranteed performance, it may be refused by the client or a deduction made from the purchase price.

From what we have just said, it might seem impossible that machines could go into operation without testing or even an hour's test run. And yet it does happen time and again; then it may be that the machine is used for quite a long time and it is only discovered that there is something wrong with it when a substantial amount of damage has been done. It will be obvious that in a case like this, the designer and his firm will be responsible for the damage.

Fig. 87.—SLM Diesel engine prepared for running tests. All important values are shown on instruments in the control room.

part 2

General Form Design

Introduction to general form design

Anyone watching a designer at work might not notice that he keeps strictly to definite rules and lines of approach. He gives the appearance of having a very wide measure of freedom in all directions, and of not being bound by set canons. It is even more surprising that the designer himself is hardly aware of working within limits and would tend to be surprised if he were asked about these rules and conditions. Nevertheless, they do exist; and an exact knowledge of them is essential for all design work. Only by keeping this fact in sight can we appreciate the pitfalls of practical engineering for anyone ignorant of these fundamental rules, who thinks he can trespass in these fields and do original work without the necessary careful preparatory study. Designing calls for special training, and we must do away once and for all with the idea that all anyone needs is a word or two of advice on how to go about things; the training is just as rigorous, in its way, as that of a mathematician.

It is sometimes said that designs can never be right or wrong in themselves, because each of them will have been arrived at with given circumstances and a given purpose in mind, and the degree of 'correctness' can only be measured against these circumstances and this purpose. To some extent this is true. The question of whether a machine bedplate, for instance, is to be cast or made up as a welded unit will depend on the production facilities and the purpose for which it is to be used. And yet it would be wrong to put things as unconditionally as that; for whether the bedplate in question is welded or cast, it will nonetheless exhibit a number of features without which it would not begin to be a usable design. These features stem from a large number of physical requirements which are encountered with every design, which are fundamental in their nature and which alter little, if at all, as time goes by—they give no hint, for example, of whether a design dates from yesterday or 50 years ago. Today, just as much as 50 years ago, seating surfaces for bolt heads still need to be milled flat, surfaces wherever a workpiece is to be machined need to be raised by a guide facing, and shaft chamfers can only be rounded and never sharp-cut. Looked at this way, there is very definitely an approach to design which is unconnected with individual requirements, which is absolute and can be right or wrong without reference to particular circumstances and a particular purpose. It is the rules of this approach to design that we are concerned with here, and it will be realized that it is hopeless to start on a design without knowing these rules fully, and observing them. Numerous though they are, the student must learn them until they are built into his memory.

In the past it was usual for the engineer to amass this knowledge through practical work, helping in the company's design office; but since this training was not as a rule tackled systematically enough, there were always gaps in the knowledge gained, and the tyro could not help making mistakes as soon as he tried to cut adrift from his instructor. This is why, even today, there is no field where 'keeping an eye' on new staff continues as far into the final stages of training as it does in the design world. Another way, however, ought to be tried today; that of giving the student this training in the basic code of design as a self-contained and systematic course.

Within this course what we have termed 'form design' will, beyond all question, take pride of place, and it is this we shall be discussing in the following chapters.

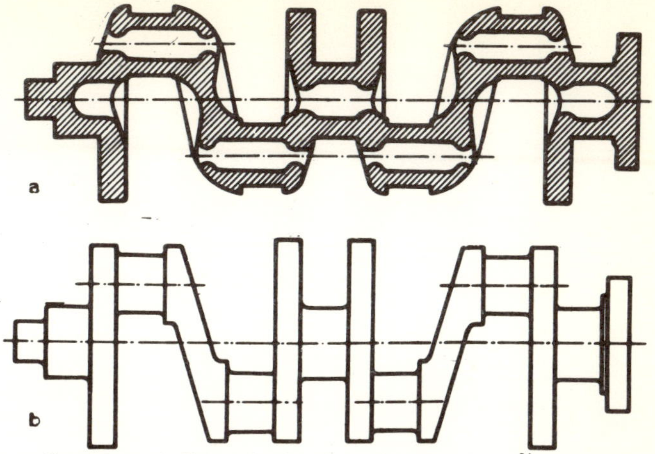

Fig. 88.—(*a*) Shape for ideal crankshaft according to Prof. Thum, compared with (*b*) normal crankshaft construction.

Elements of design

The whole of the engineer's design work is carried out under a strain which it would not be an exaggeration to call tragic. The strain is evident in everything Man does, and comes from the insuperable discrepancy that always exists between what we would like to produce and what we are able to produce.

Most of the physical processes in our machines follow complex laws which can never be exactly reproduced and can only be shown in an approximate form by lengthy mathematical formulae. Cases where these laws can be stated by a linear, quadratic or trigonometric function are very few and far between. The parts of a machine would be just as complicated if we tried to match them exactly to the laws, and we would fail in the attempt because of the lack of any mechanical production process for making them economically. We now know, for example, what form an ideal crankshaft should take to be specially resistant to bending and torsional loads (Fig.

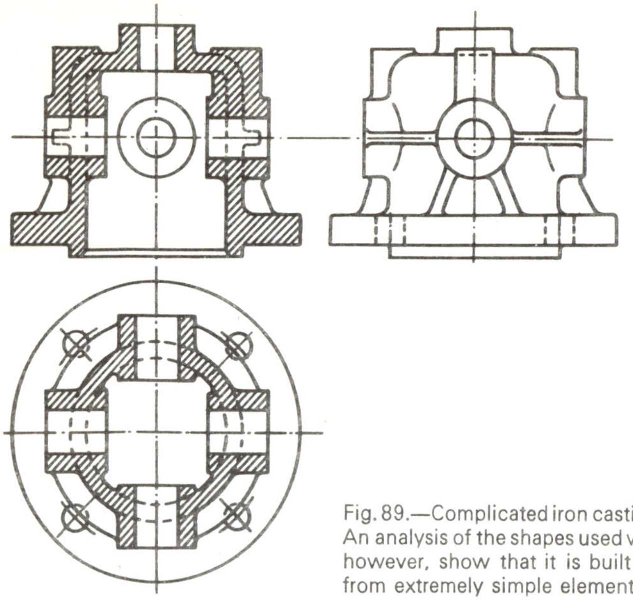

Fig. 89.—Complicated iron casting. An analysis of the shapes used will, however, show that it is built up from extremely simple elements.

shop drawings; parabolas and hyperbolas are seldom seen, and curves of a higher order even less. Helicals, on the other hand, are to be found since they can be very simply compounded from motions of translation and rotation. A drastic example of how manufacturers prefer simple geometrical forms is seen in the blades of small fans; these are usually made from sheet metal of uniform thickness bent into part of a cylinder, although this shape is far from ideal on aerodynamic grounds (Fig. 91).

It is also striking how often the straight lines seen on engineering drawings are at right angles to each other, and we can only conclude from this that the vertical and horizontal are the most favoured directions in design. This is partly so because there is frequently a connection between these directions and the method of functioning; furthermore, such right-angled junctions are deliberately preferred to those in which a part is joined at a sloping angle, where the insertion and transition of surfaces is a good deal more troublesome.

These comments should be enough to show that in solving his problems the designer is always faced with two contra-

Fig. 90.—Components of a +GF+ (George Fischer) ball cock in plastics, showing the extreme simplicity of form of its elements.

Fig. 91.—GM (General Motors) vehicle Diesel engine. The fan is an example of how simple shapes are preferred for production reasons, even though they may not be the most efficient shapes.

88*a*); the only trouble is that it is impossible to produce such shapes by pressing or forging. They could, admittedly, be cast: but this would mean sacrificing the highest quality of materials, and is unacceptable for this reason. And so we set this design aside, and carry on making crankshafts as shown in Fig. 88*b*. Metal machining processes are based almost without exception on simple types of movement such as translation and rotation, and these only allow economical production or correspondingly simple forms such as sheets, prisms, cylinders or cones. This is why normal constructions are made up of such simple component parts (Figs. 89 and 90). In just the same way, circles and straight lines predominate in work-

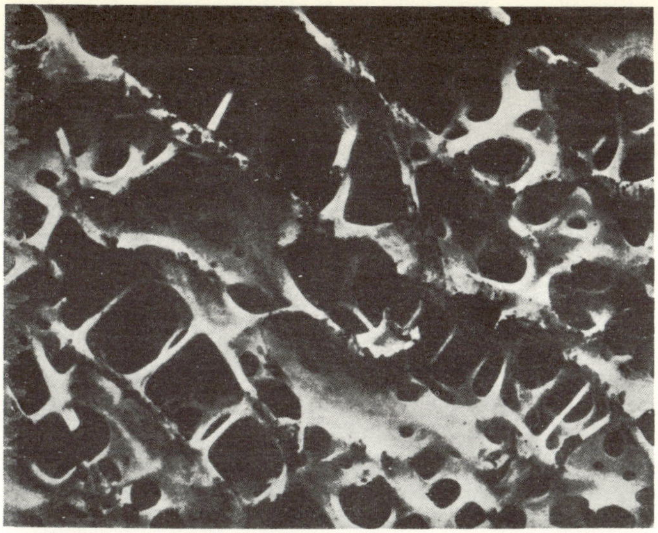

Fig. 92.—Part of the load-bearing structure of the human hip bone. It can be seen that there is no single part which is either circular or straight.

Fig. 93.—The chambered nautilus. The creature lives in the first open compartment, remaining fixed to the centre of the shell by a cord passing through all the others. Each year it forms a new outer compartment, into which it 'moves'.

dictory requirements: the first is to arrange his designs so that their functions are carried out in the best possible way, and the other is to arrange them so that they can be produced by straightforward simple forming processes and machinery. The designer has to find a compromise, and to make enough concessions in each direction for the result to be an optimum in the particular circumstances.

In many things Nature can teach us a great deal about constructional skills, and the designer would be well advised to follow her precepts. He will, of course, find that in choosing what shape to use she is not subject to the same restrictions and limitations as he is—she imperiously obeys the laws of life itself, producing forms which sometimes inspire our greatest admiration (Figs. 92, 93 and 94). Nor are Nature's workshop drawings dominated by the circle—the shapes in the majority of cases follow laws which it would be quite impossible to put

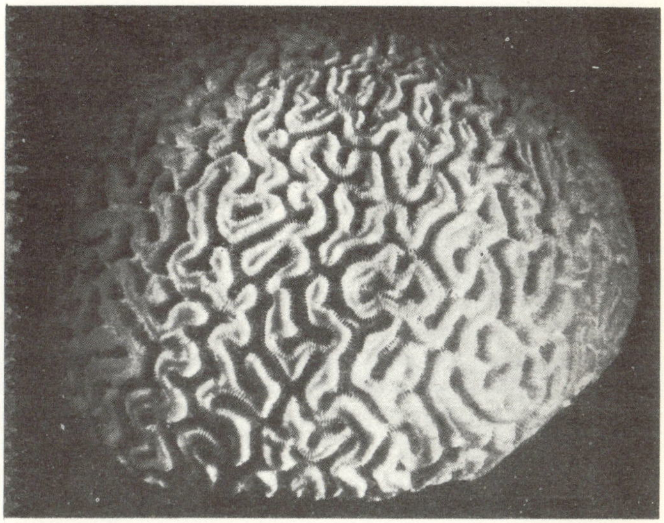

Fig. 94.—Base for coral, carrying a large colony. The skeleton is extremely well matched to the living habits of the minute creatures.

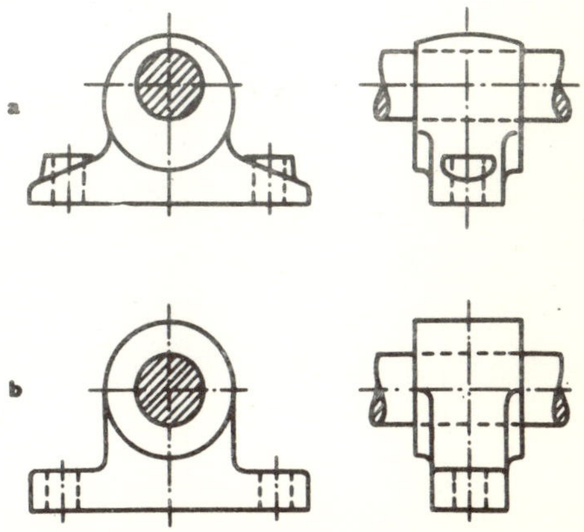

Fig. 95.—Small pedestal bearing, with optimum constructional form for (*a*) strength and (*b*) simplicity of manufacture.

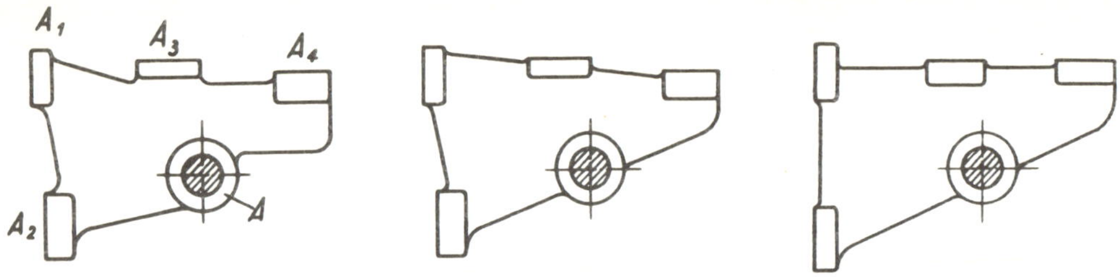

(*a*) first sketch

(*b*) simplified basic form

(*c*) with slightly modified specification for the bearing surfaces A.

Fig. 96.—Simple bracket as an example of step-by-step improvements in layout and simplicity.

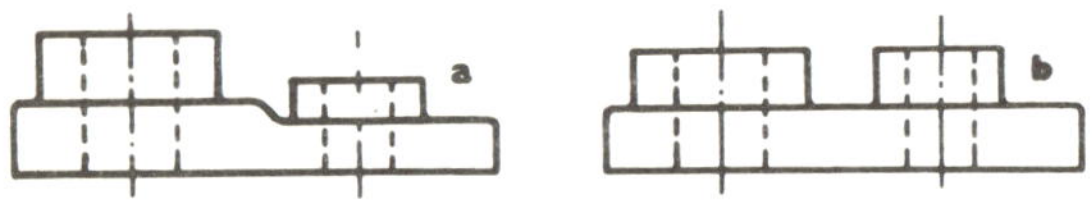

Fig. 97.—Small bearing plates, with ideal form for (*a*) function, (*b*) production.

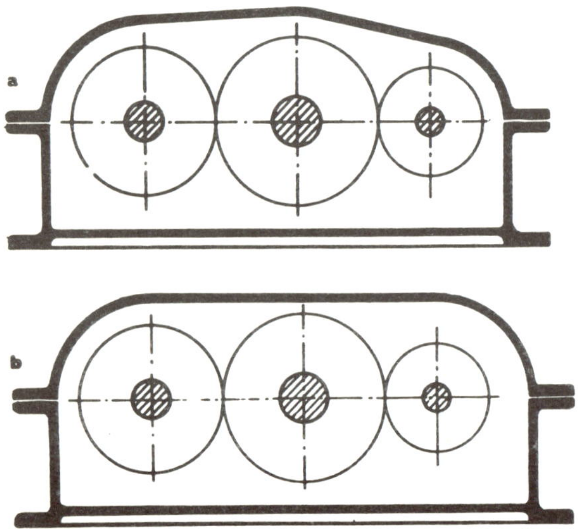

Fig. 98.—Gearbox casing:
(*a*) matching the size of individual gearwheels
(*b*) preferable for ease of manufacture and neater appearance.

in the form of an equation at all. We can be quite safe in saying that our machines would look totally different if, in making them, we had anything like the opportunities that are available to Nature. It would, for instance, be very reasonable to give a pedestal bearing the form shown in Fig. 95*a* from the point of view of distribution of forces; but because we have to think of the difficulties entailed in producing it, we alter the shape to that seen in Fig. 95*b*. So all we can do is to take over Nature's examples of how to go about things, and follow our own tortuous paths in putting them into effect.

The basic feature of successful designs is invariably simplicity and clarity. If we consciously try to achieve these, it will make almost everything connected with manufacture much easier. To start with, the passing of information and instructions to the place where the work is done, and the means generally used for this—the workshop drawing—will be made simpler. Every dimensional indication that can be saved will help; there have been instances where technically excellent designs have been rejected because the difficulty of making them understood by all was considered to be insuperable. The simplicity we are talking about here has to be sought —it will not come about by itself. It may happen that the initial draft for a design seems to offer an entirely practicable solution to the design problem. Critical assessment may show,

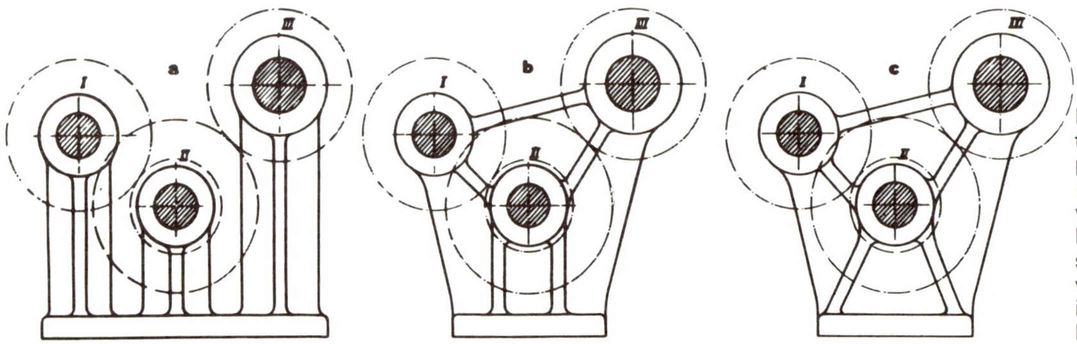

Fig. 99.—Group of three bearings supported on baseplate: (*a*) oversimplification, with each bearing linked to baseplate separately, (*b*) and (*c*) with crosslinks matching the mechanical linkage of the gears.

however, that the whole thing is still not a unity because, in considering the various needs that have to be satisfied, too much attention has been paid to details without taking an overall view. Things should not be allowed to rest here—an attempt must be made to bring the various points together in a simple and clear overall concept. Fig. 96 shows a design job where a bracket has to be made which will match up with the existing parts of a machine at points A_1, A_2, etc. Fig. 96*a* illustrates a first attempt, the fussy shape of which is difficult to keep in the mind's eye and for which the quoting of dimensions would be a complicated business; for these reasons it is found unsatisfactory. A slight effort at simplification results in the version seen in Fig. 96*b* which already has a very much clearer line. As this is still not good enough for us, we shall try to obtain a slight modification in one or two points of the specification; sometimes we shall not succeed when we try this, but very often slight alterations are possible. In this case, it would be a matter of moving surfaces A_1 and A_2 so that they share the same vertical, and surfaces A_3 and A_4 into the same horizontal; the result of this is seen in Fig. 96*c*. The outcome of these modifications is that the shape of the bracket—a trapezoid—is easy to remember and the total number of dimensional indications needed is about 15 instead of 30.

Substantial improvements can often be achieved in this way by quite small modifications, and this can again be seen in the example illustrated by Fig. 97. When dealing with gear trains we have a well-tried rule of keeping the distance between the individual gearwheels and the casing as constant as possible; but in the case seen in Fig. 98 this would lead to an ugly and asymmetric shape for the housing (Fig. 98*a*), and so we have to consider whether in this instance it may not be better to ignore this rule for the sake of appearance, and to adopt the simpler shape seen in Fig. 98*b*.

This arranging and changing must not be misunderstood, however, and must not result in a simplification which runs counter to the fundamental relationships of the mechanism concerned. This can be seen happening in Fig. 99, where three shaft bearings have to be mounted on a common bedplate. One way of dealing with the problem would be version *a*, where each bearing is joined to the bedplate separately, but this would be ignoring the important fact that there is to be a mutual transfer of forces between the shafts via gears, in the course of which the individual loads on the bearings—forming the action and reaction for one and the same force—will act outwards, so that there is no need to pass these loads to the bedplate at all. Consequently, it would be better to allow the transfer of forces to balance itself out by linking the various bearings along the lines shown in versions *b* or *c*. The rightness of this approach is confirmed perhaps by the fact that these latter solutions strike the eye as being better than solution *a*.

The principle of constant wall thickness

The aim to make everything as simple as possible for the workshop leads to a basic rule of design—the principle of constant wall thickness. This means that hollow parts and flat or ribbed workpieces produced by casting or welding should preferably be made with a given uniform wall thickness. Seen from the functional viewpoint, there is in fact everything to be said against such a rule—take, for example, the strength characteristics of the pressure vessel seen in Fig. 100. According to the 'boiler formula' this ought to be made with a number of differing thicknesses δ, matching the diameter *d* at various points. Since $d_1 < d_2 < d_3$, we should make $\delta_1 < \delta_2 < \delta_3$. The flat base to the left is a case of its own, and ought according to theoretical mechanics to be especially thick. In spite of these reasons for varying the thickness of the wall, the container will probably be made with a uniform wall thickness, apart from the flanges which are particularly heavily stressed. Just how natural it is for a designer to keep to the same wall thickness can be seen from Fig. 101; here every section of wall ought properly, from the functional viewpoint, to have a different thickness—and yet more or less the same thickness has been used throughout.

There are five main reasons for following such a rule:

1. First of all, there is the question of sheer convenience which prompts the designer to use the same wall thickness wherever possible.
2. The working drawings are simpler, since the wall thickness needs to be indicated only once. There is no possibility of confusion arising, and the craftsman finds it easier to remember one figure than a whole series.
3. When castings are made with a uniform wall thickness, the cooling is more even, fewer heat-stresses are set up, and the risk of shrink holes forming is reduced.
4. With castings the strength of the workpiece is more uniform, because this is to a large extent governed by the rate of cooling. The more quickly a casting cools, the greater its strength; so the thinner the wall, the greater the strength of the casting.
5. When parts are to be fabricated by welding sheet metal and angle iron, maintaining stocks of these basic items is simplified.

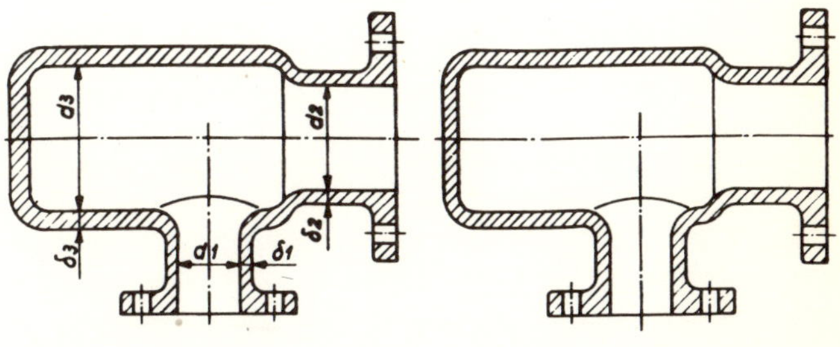

Fig. 100.—Pressure vessel, left, in which the wall thickness matches the varying diameter. In production, however, the version on the right with constant wall thickness was used.

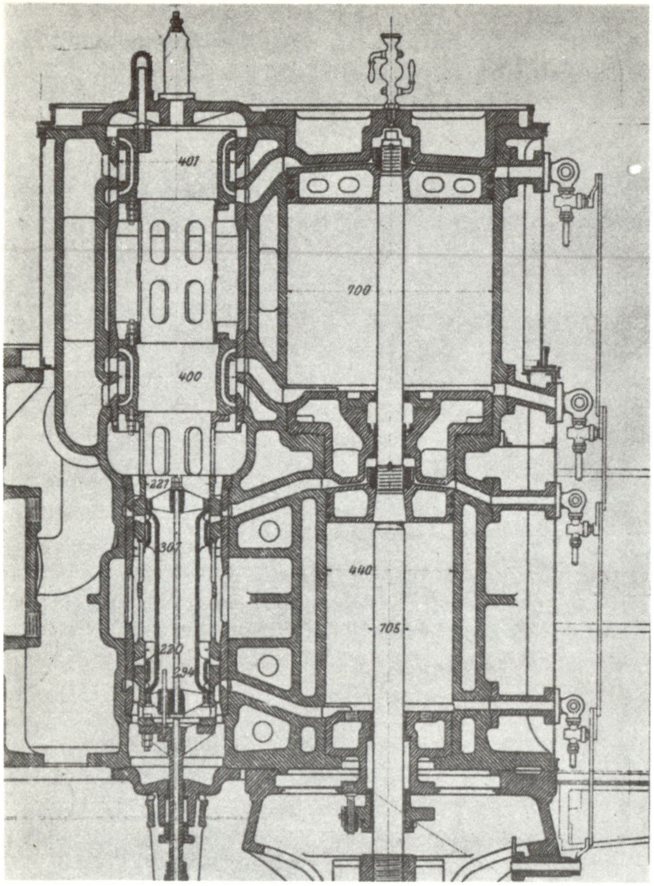

Fig. 101.—Cylinder and slide-valve housing in a tandem engine, with almost entirely uniform wall thickness, even though varying thickness would meet the strength requirements better.

There are many examples in design practice which show how uniformity of wall thickness is adopted in the face of the requirements of functional efficiency. The first of these we might mention is the everyday lever. Here, as we know, the greatest bending moment occurs close to the pivot end, and it would thus be quite reasonable to make the cross-section at this point both wider and thicker, like the version seen in Fig. 102*a*. It is, however, equally possible to allow for the greater bending moment by giving the lever the same thickness throughout and increasing the width at the cross-section still further (Fig. 102*b*). The same considerations lead, in the design for a wheel, to the thickness of the wall and arms being kept the same on the inside and outside, resulting in the form shown in Fig. 103.

A typical example of how a variation in wall thickness is avoided can be seen in the cast-iron pressure plate, made to take two laterally-acting forces, in Fig. 104. It does not need much thought to conclude that the centre part of this plate is lightly stressed and that a fraction of the general wall thickness

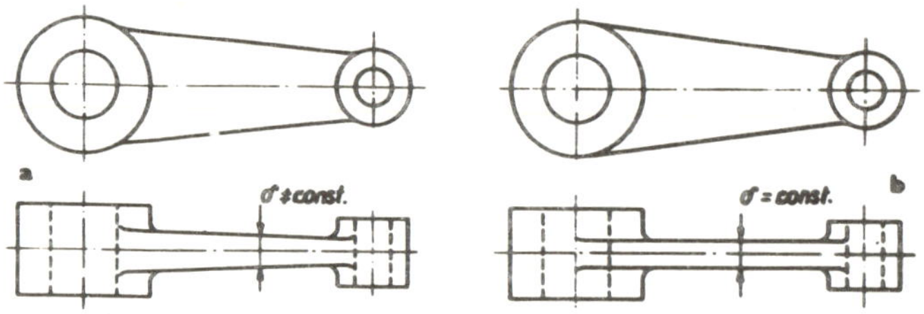

Fig. 102.—Levers. The bending moment decreases outwards, and this variation can be dealt with by varying the cross-section thickness. However, the version on the right, with constant thickness, is preferable.

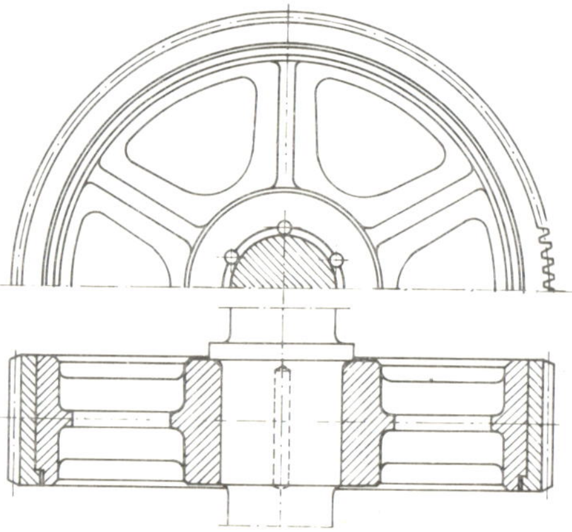

Fig. 103.—Wheel construction, again following the principle of constant thickness.

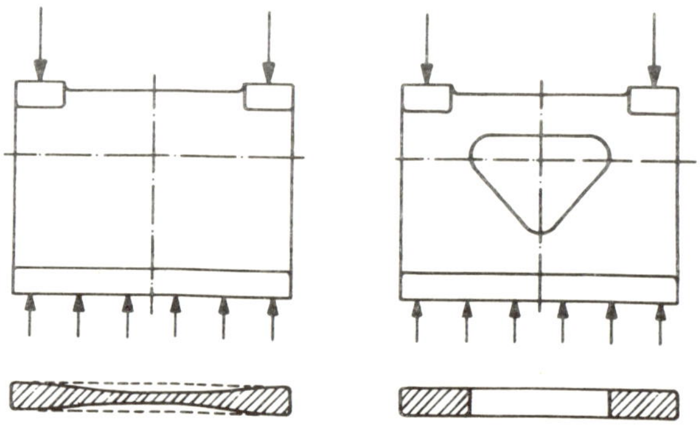

Fig. 104.—Pressure plate—left, with a thinner section at the area of low load, and right, with a lightening hole.

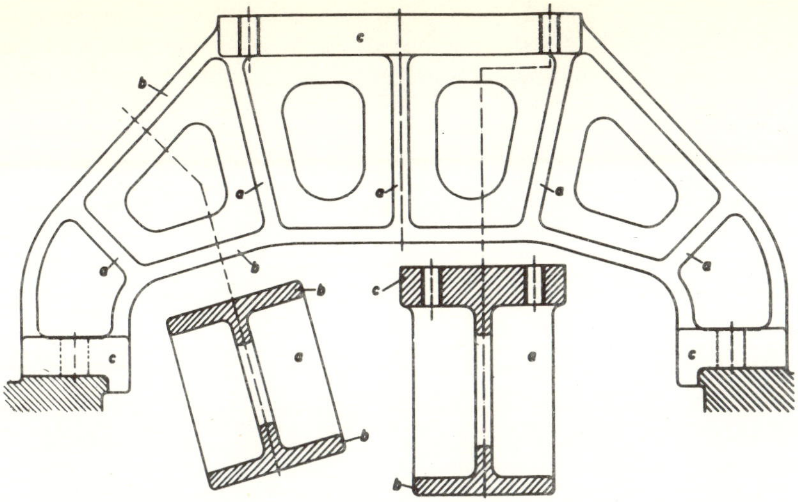

Fig. 105.—Bearing lock for a centrifuge. Areas of low load have been given lightening holes, and ribs *a* have been added between them to stiffen the flanges *b*.

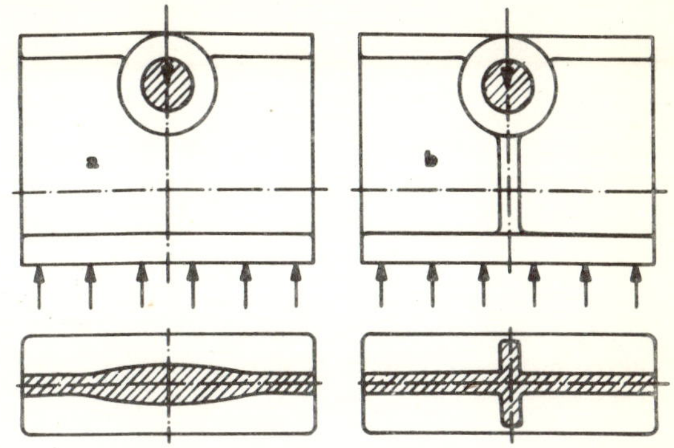

Fig. 108.—Support plate with cast-in bearing: (*a*) with high-load area thickened, (*b*) with strengthening ribs.

used would suffice to cope with the loads that occur. Nevertheless, the plate will be made with the same thickness everywhere, and at areas of very low stress some of the material will simply be omitted altogether. This gives a 'lightening hole' (so called although its primary purpose may not necessarily be to make the workpiece lighter), a feature encountered in a great number of designs where there are lightly-stressed areas (Fig. 105). There is, however, a danger involved in these lightening holes: they represent a break in the force transfer pattern, so that an increase in stress along the edges must be allowed for. In many designs, in fact, cracks have appeared at these places, and it may sometimes be a good idea to strengthen them by adding a small edge in the form of a flange (Fig. 106). Bare edges on a part are always rather vulnerable, in any case, and should be finished off with a flange. All reinforcements of this kind will, of course, count as parts of the wall and must therefore be of the same thickness. They should, moreover, have a simple shape; Fig. 107 shows four examples, of which the fourth is preferable to the first three.

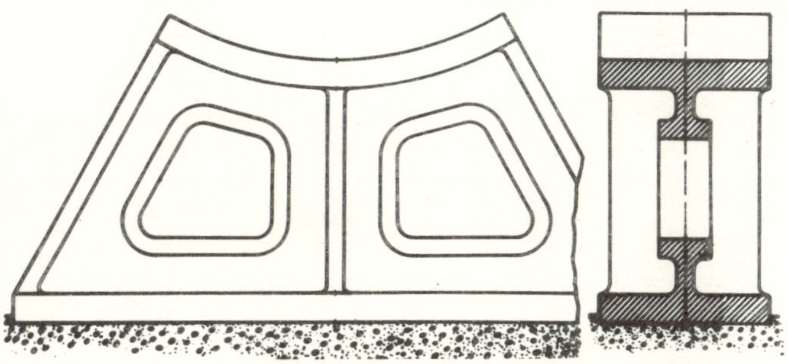

Fig. 106.—Pipe saddle, with lightening holes in the centre wall which are flanged at their edges to prevent cracking.

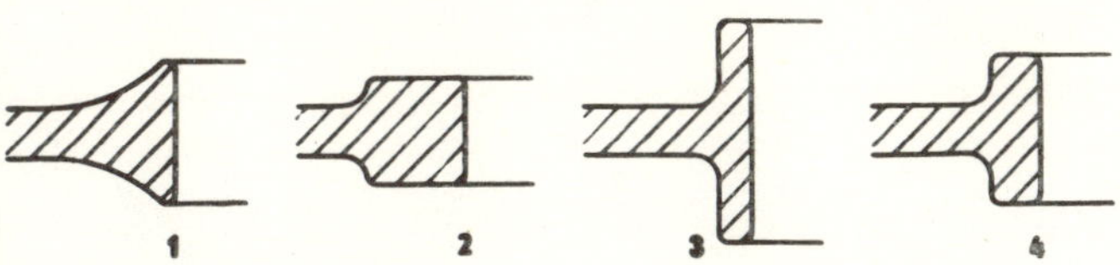

Fig. 107.—Various forms of edge reinforcement: (1) too complicated, (2) too massive, (3) too thin, (4) simplest form with same thickness as wall.

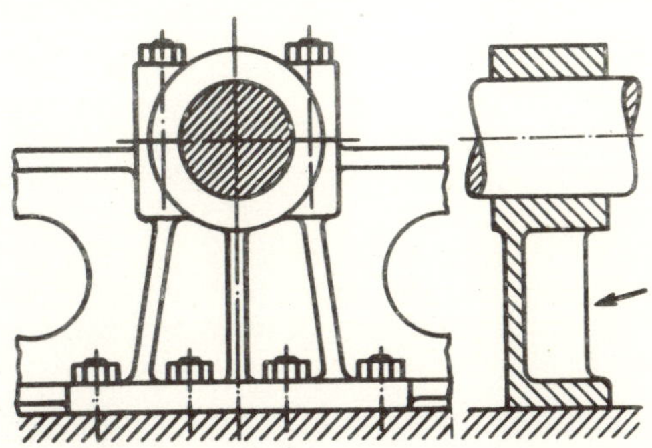

Fig. 109.—High bearing loads call for extra ribs (arrowed) to strengthen the wall beneath the bearing.

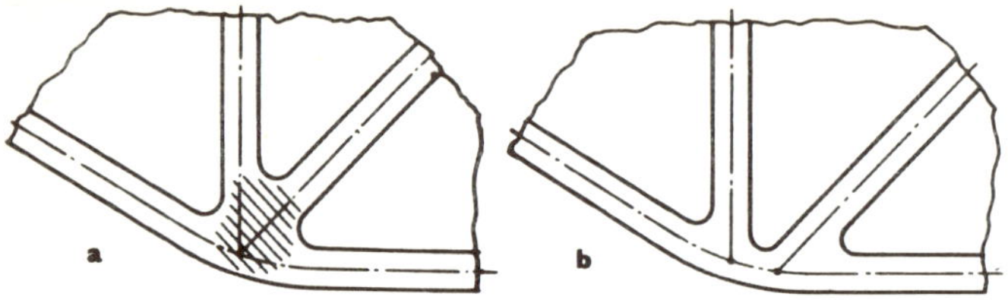

Fig. 110.—Joints of ribs:
(*a*) convergence at one point means a concentration of material
(*b*) slight offsetting of the rib axes avoids this.

Fig. 108 illustrates a case which is in some ways the very opposite of what we have just been discussing: instead of there being a reduction in the amount of force being transferred through the centre area, there is an increase. One answer to this might be to correct in the opposite direction, and make the material thicker at this point. But this is not done; instead a rib is added to strengthen the wall. All that the extra loading at this point demands is a greater cross-sectional area; consequently we are free to choose the profile for the rib, and we again give it the same thickness as the wall itself. The depth of the rib will be adjusted to give the extra cross-sectional area required.

If the extra load exerted on a wall at one point is very large, two or three ribs will be used in place of one very deep one. Even when there are several ribs, they will all, of course, have the same thickness as the wall itself (Fig. 109).

A special problem arises when several ribs come together to meet at a single point, as in Fig. 110. Unless we are careful, we shall get a thickening at this point (Fig. 110*a*) which is undesirable from the casting viewpoint since it will prevent even cooling. Nodes of this kind can, however, be avoided quite easily by offsetting the ribs somewhat from the point of junction, as shown in Fig. 110*b*.

Ribs always (provided they are really being used for strengthening and not just for stiffening) lead to other force-carrying parts, as can be seen, for instance, in Fig. 109. If we think ahead in such cases, we shall not bring the rib out to the very edge of the adjacent part, but will set it back slightly (arrowed in Fig. 109). For if subsequent machining work on

Fig. 111.—Sulzer centrifugal pump. The ribs in the feet are the same width as the sole plates, as no machine finishing is needed.

Fig. 112.—Projecting edges on casings should, as a general rule, have a small radius. In *a*, however, this thickens the wall, so the rule is abandoned and the external radius enlarged as in *b* to avoid the thickening.

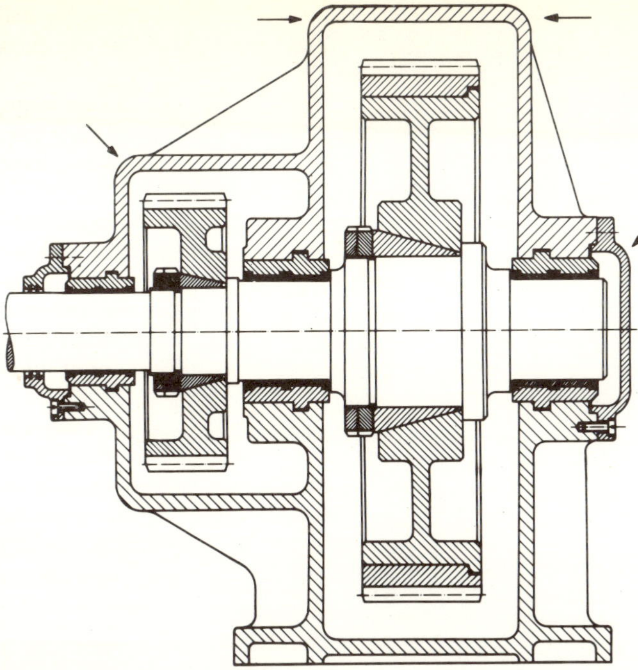

Fig. 113.—Gear housing as an example of edges rounded off as shown in Fig. 112.

these parts should be necessary (say, because the workpiece is a casting), then the edges of the ribs would have to be machined as well—which could be both troublesome and costly. When there is no likelihood of this happening, however, the ribs can be brought right out to the edge (Fig. 111). The same applies to welded constructions, where machine-finishing of this kind is seldom needed.

The principle of constant wall thickness is usually chosen too in cases where it competes with other rules aimed at simplifying the design—in the shaping of the corners, of casings and containers, for example. Here there is a rule which says that the projecting edges of bodies should be given a very small radius of curvature. If this is in fact done, we will end up with a shape like that seen in Fig. 112*a*, with a local thickening that will be frowned on in the foundry; and so in this case we shall abandon this second rule in favour of the first and round off the corner with a gentle enough curve for the inner and outer arcs to have the same centrepoint and for there to be no thickening of the wall (Fig. 112*b*). We can see this in practice, in Fig. 113, in the large radii used for the rounded edges on a gear housing.

This principle of constant thickness is not confined to one and the same workpiece, but often extends to all the parts that are assembled to form a unit. This goes without question in the

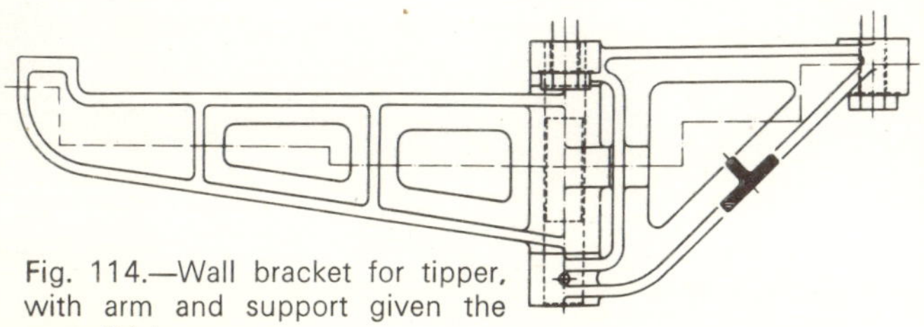

Fig. 114.—Wall bracket for tipper, with arm and support given the same thickness.

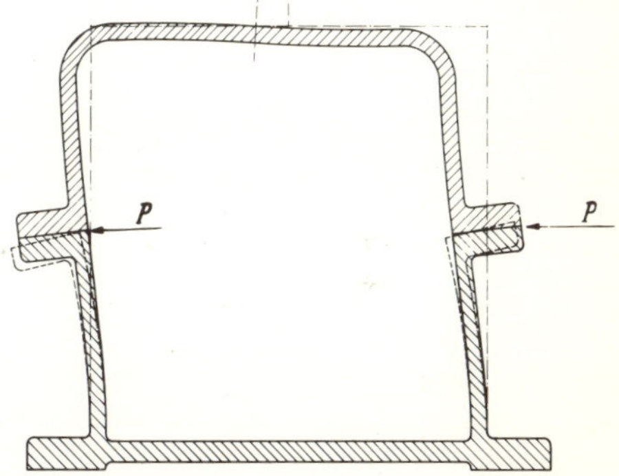

Fig. 115.—Distortion of gear housing under lateral load when top and bottom sections are bolted together and have the same wall thickness (dotted lines show distortion of the lower section when this alone carries all the load).

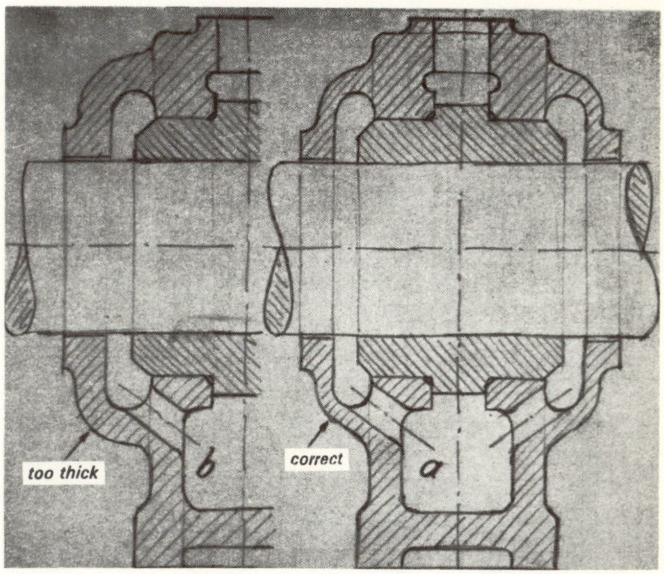

Fig. 116.—Bearing housing as an example of the occasional necessity of varying the wall thickness. Version *b* shows clearly the pointlessness of keeping to the same thickness, compared to *a* (freehand drawing).

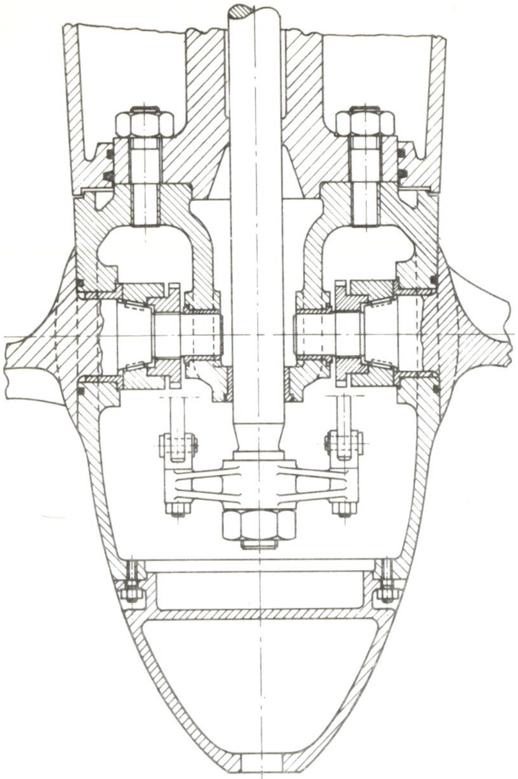

Fig. 117.—Kaplan turbine hub: in the present context, it serves as an example of low stress on the lower part of the housing, as the forces from the blade shafts pass upwards into the flange. Wall thickness in the lower section can be reduced.

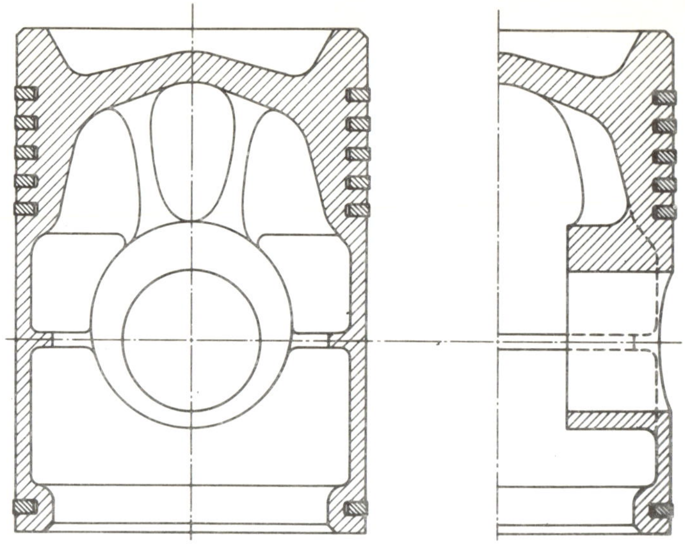

Fig. 118.—Internal-combustion engine piston. The skirt is only stressed through acting as a guide, as the main load from the exploding gas passes from the piston crown via the gudgeon pin to the connecting rod. The skirt can therefore be much thinner.

case of a divided bearing where the upper and lower parts are, of course, made the same thickness (Fig. 109), but it is also applied where there is not such a close unity, as in the case of the side-walled bracket of Fig. 114. Less straightforward is

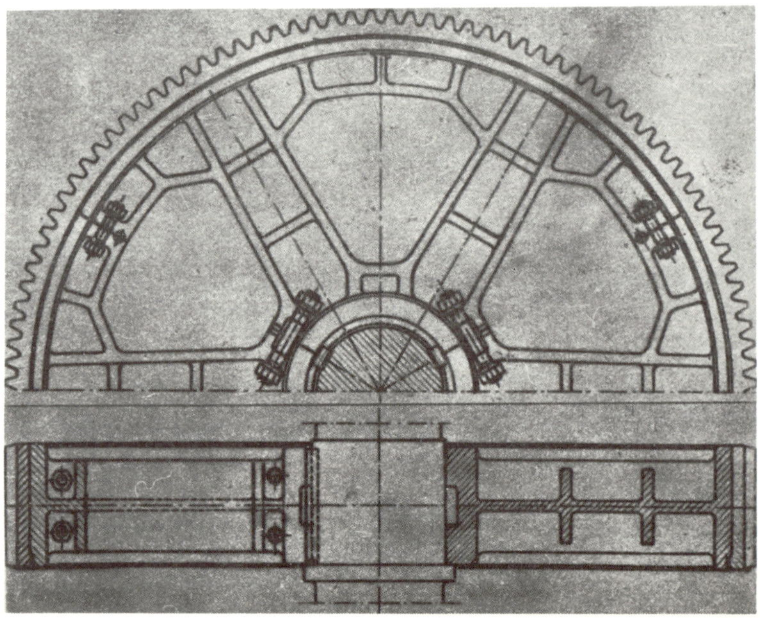

Fig. 119.—Multi-section gearwheel: cross ribs in the arms and ring are an example of load-free supports intended for stiffening only, and they have a slightly reduced thickness.

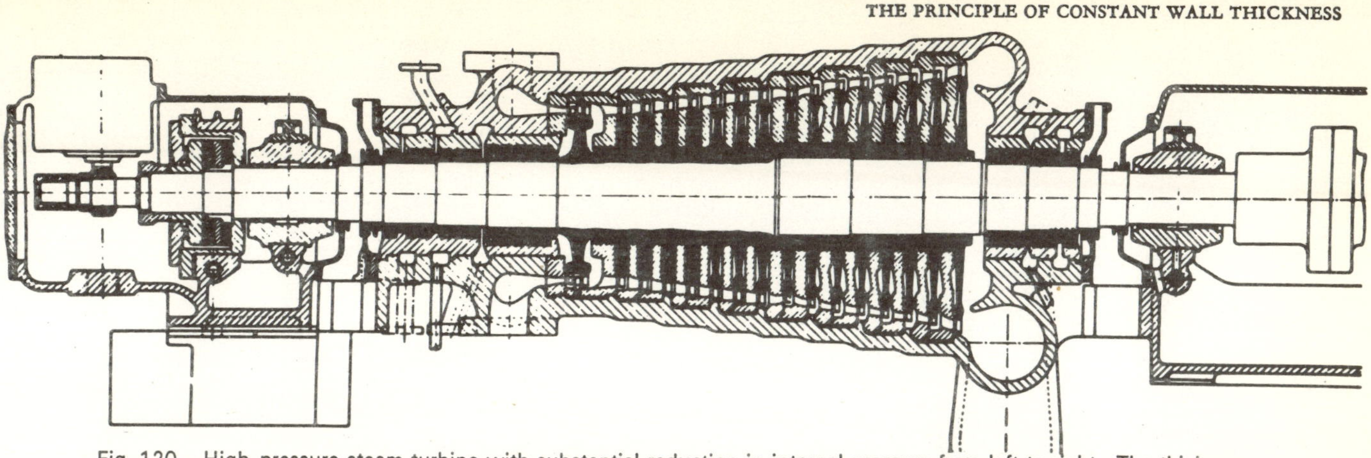

Fig. 120.—High-pressure steam turbine with substantial reduction in internal pressure from left to right. The thickness of the wall is matched to the pressure, i.e. is progressively thinner from left to right (Escher Wyss, Zurich).

the case of the two-section gear housing of Fig. 113, where the view is sometimes held that the upper section is no more than a protective covering for the internal parts, so that it takes no mechanical loads and can be given a thinner wall than the bottom section. This is not, however, true (or is true only to a limited extent) since, as they are bolted together at the mid-level, the upper and lower sections form a whole from the point of view of mechanical stresses, and are exposed to these to an equal extent (Fig. 115). The right answer is therefore to make both halves with the same wall thickness, as has been done with the gear housing of Fig. 113.

There are, of course, exceptions to this law of constant wall thickness, for it would be simplifying things too far to give it more than general importance as a guide. There are many circumstances in design work where it becomes necessary to vary the wall thickness in certain areas, and where neither lightening holes nor strengthening ribs will meet the case. This applies, in the first instance, to cases where the loading on a part of the wall not only decreases but in fact disappears altogether, and yet the wall has to be retained in order to enclose a space or provide protection. We can see an example of this in the bearing housing in Fig. 116*a*, where the force transferred from the shaft to the bearing is entirely led away to the bedplate via the two vertical supporting walls, and the oil-traps at the sides are unstressed. Apart from knocks during assembly or transport, there is no force exerted on them at all; they are, however, indispensable for catching the lubricating oil thrown out sideways. It would be quite pointless to give them the same thickness as the parts carrying the bearings—this would be an obvious infringement of the law of economical use of labour and materials from which, after all, the law of constant wall thickness is derived. If the tyro designer should doubt that the right answer in these circumstances is to make the casing thinner, he has only to look at the other way of doing things shown in Fig. 116*b*, where the wall is the same thickness throughout. We also have to consider the difficulties presented by the casting of a component like this, and must try to make the transition between these two very different parts of the wall as gentle as possible. First, a large curve will be used, but at the same time a gently increasing thickness of the wall will be provided between the two extremes. Both features are combined in the unit seen in Fig. 116*a*. The foundry people, too, will use their own methods of preventing casting stresses and the formation of shrinkage cavities, by inserting chills in the moulds when casting parts with very thick walls. It will, of course, be for the designer to decide where he is going to let the foundry adopt such measures, which will naturally make production more expensive.

Further examples of this kind are shown in Figs. 117 and 118. In both these designs the total loading leaves the lower section unaffected; but these have a part to play—serving as a water passage and guiding the piston respectively—so they cannot be left out of the designs.

In general a rib is provided for strengthening purposes; but there may be cases where it is intended solely as a safeguard against possible tilting or buckling. With I-beam sections the wide flanges have a tendency to fold inwards when under compression, so ribs are added at regular intervals. These are crosswise to the direction of the force, so they carry no load, and serve only to stabilize the position of the flanges. As they carry no load, it would be overdoing things to make them the same thickness as the adjacent wall; from the casting viewpoint, too, there is no objection to giving these ribs 70 or 80 per cent of the general wall thickness (Fig. 119). The casings of turbine machinery such as pumps, compressors and steam turbines are another example of varying loads; the processes taking place within them involve considerable changes in pressure, and there may often be a very large pressure difference

Fig. 121.—Exhaust chamber of a large steam turbine: the suspension bosses represent an exception to the rule of constant wall thickness, as does the support flange (top left); in spite of the ribbing, these points are made much thicker (*Photo: Brown Boveri, Baden*).

between the input and output side. Such casings are now no longer made with a constant wall thickness, which is matched instead to the pressure in the various sections. This is all the more reasonable since the wall thickness changes from one extreme to the other over quite a long distance, so that casting problems are not likely to arise (Fig. 120).

A further group of exceptions covers cases where there is a massive concentration of force flow at individual points. Typical examples of this are the suspension points provided on vessels and housings (Fig. 121), as well as seatings and points of attachment to adjacent parts (Fig. 122). Reinforcement is needed at all these points to take the load concentration and then distribute it over the neighbouring areas. This is one of the reasons for the typical configuration of gearwheels, in which the centre hub and peripheral ring are always thicker than the radial arms (Fig. 119).

Since bolts also serve to transfer forces from one part to the next, reinforcement is needed around their seatings, just as it is at points carrying axles and shafts (Fig. 123). We can see how essential this sort of reinforcement is from the consistency with which it has been applied at all doubtful points of the bearing housing seen in Fig. 124. As this illustration shows, the reinforcement involved is often quite massive, so that it is reasonable to wonder whether in instances like this we ought not to relieve the stress by adding ribs or providing more than one wall. Fig. 125 illustrates a flanged bearing supporting the large shaft of an agitator. It would hardly be permissible to bolt the heavy flange ring direct onto the unreinforced wall of the housing (far left), so in the area affected by the bearing this wall is made thicker and strengthened by means of radial

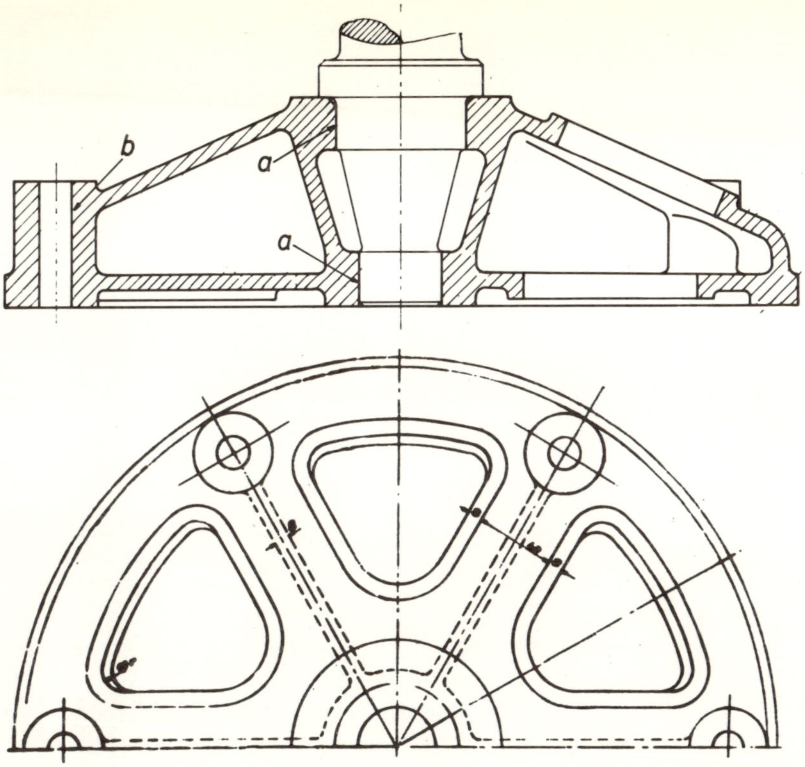

Fig. 122.—Bedplate for a rotating-jib crane. The bearing points *a* for the pillar are shaped as very thick central rings to withstand the high lateral pressures; similarly the sides of the holes at *b*, which take the anchor bolts, are made very much thicker.

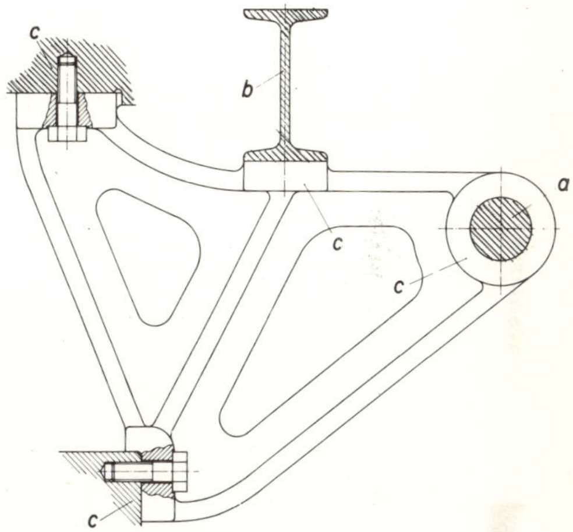

Fig. 123.—Small bracket carrying a shaft *a* and beam *b*, with carefully reinforced areas at *c* for coping with the loads involved.

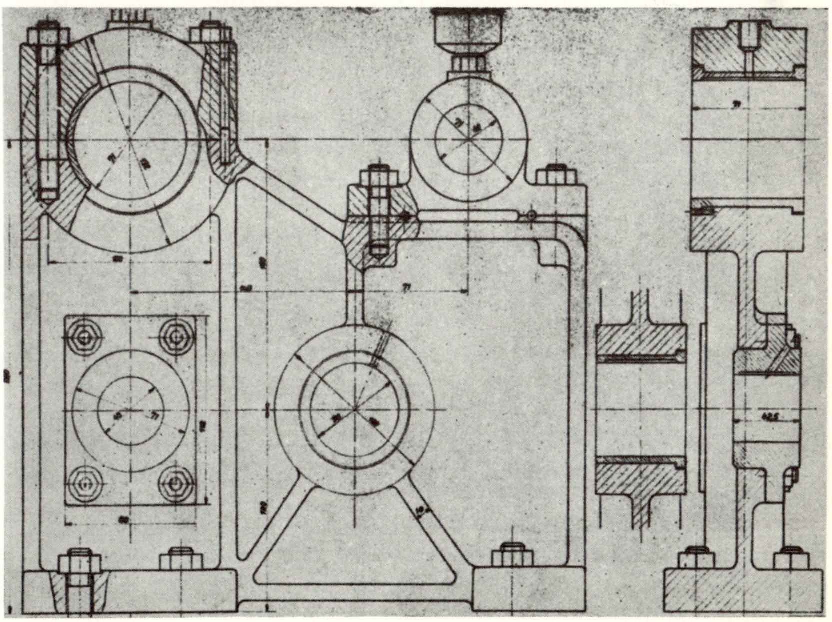

Fig. 124.—Bearing housing for lift gear, as a simple cast construction. All points carrying a load—bearings and feet—are given extra thickness.

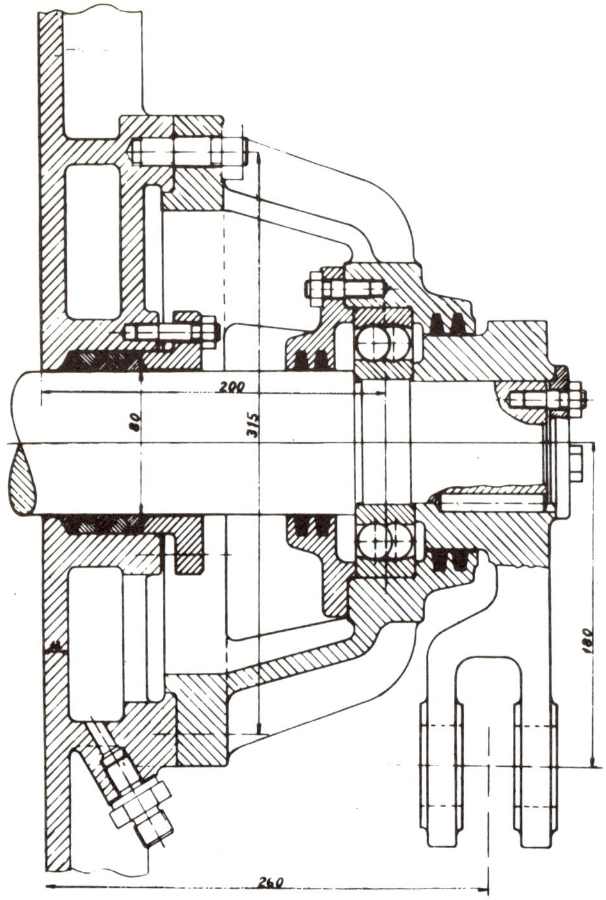

Fig. 125.—Agitator paddle drive with high loading. The wall thickness of the housing is doubled in the areas affected by the shaft load, and further strengthened by radial ribbing.

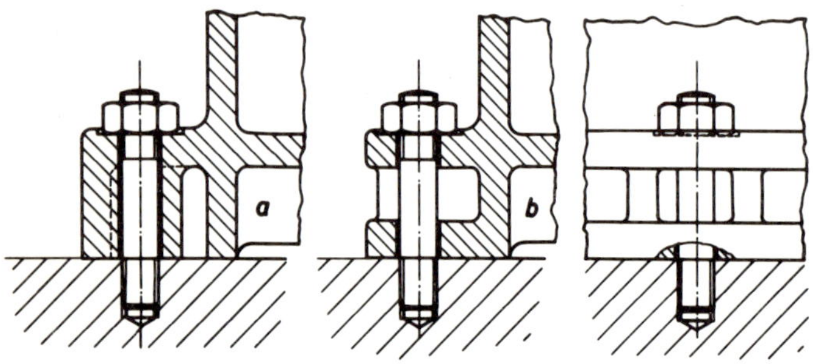

Fig. 126.—Skirting of a housing with substantial reinforcement to allow for bending loads. In order to avoid casting stresses, the skirting is made hollow (left) or given double walls (right) with internal ribs.

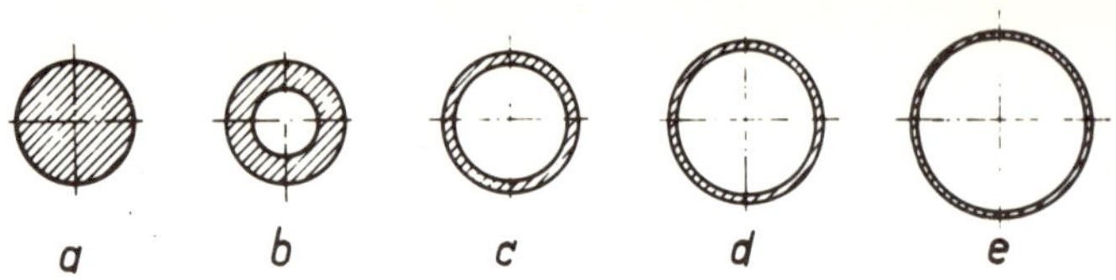

Fig. 127.—Comparison of sections with equal second moments of area and varying wall thickness. The improved utilization of materials as the wall thickness is reduced is obvious. The cross-sectional areas are in a ratio of 1 : 0·72 : 0·47 : 0·32 : 0·22.

ribs. The same sort of thing can be seen in the skirtings of heavy housings (Fig. 121). When these are castings, the skirtings are made hollow (Fig. 126*a*) or given a profile open at one side (Fig. 126*b*).

So we see that most designs represent a compromise between obedience to the rules and occasional well-founded deviations from them. Yet it is this careful mixture that gives many projects the individual features that mark them as mature and sophisticated designs.

Strong and weak shapes

When designs involve a high level of mechanical stress, the way in which the material is arranged is of immense importance. If a bar is intended to take only a tensile load, it can equally well be of round, square, or oblong section—all, provided they are of equivalent cross-sectional area, offer the same strength. If, however, the tension is replaced by compression, the shape of the cross-section is immediately of importance; a bar may buckle under compression, and the closer the material is to the centroid of the cross-section, the easier it is for this to

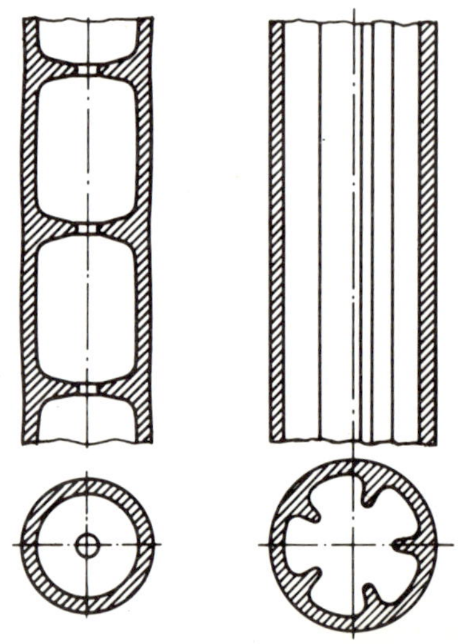

Fig. 128.—Constructional forms used by Nature: stalks with annular ribs (as in rushes and bamboo) or longitudinal ribs (grasses, etc.).

Fig. 129.—Arched bridge to carry a pipeline, itself fabricated from tubing.

happen. Buckling is a form of bending in which the decisive factor is the second moment of area (sometimes called 'moment of inertia') and, depending on the length of the bar and the load to be carried, there is a given value below which this quantity must not be allowed to fall. Since the buckling may occur in any of the planes, the second moment of area should ideally be the same for all cross-sectional axes. This condition is met by a round tube, as well as by a square box girder. The further outwards the cross-sectional area is located, the greater the saving in material (Fig. 127). This does, however, introduce a fresh danger—that of bulging (the walls forming the round or square profile may bulge under load if they are made too thin). Nature provides us with shrewd examples of how to get over this difficulty. First of all, she uses for compression members almost exclusively a closed hollow profile (e.g. hollow bones). In the stalks of tall grass she shows us how to overcome the danger of bulging, by using either ring-shaped ribs at regular intervals on the inside—these even meet sometimes to form a complete 'floor'—or ribs running the length of the walls (Fig. 128). These examples can well be borne in mind when designing compression members; with small sections such as are common in frameworks this method is not practicable, although it is quite possible when the diameter of the work leaves enough space inside the support.

There has been a great increase in recent years in the use of tubing and other closed sections as compression members (Figs. 129 and 130), and their superiority over any kind of open section seems to have been generally recognized. For the same second moment of area, closed sections are less liable to bulge than open ones, which in the past have been used almost exclusively for compression members (Fig. 131); open sections tend to fold at the outer parts of their profile when under compression (in the direction indicated by the arrows in Fig. 131), thus lowering the second moment. This can be resisted by adding stiffening webs at the danger points.

Special conditions obtain where there is torsional loading, and here innumerable sins are committed by the designer. In many instances he does not fully understand the inherent nature of torsional loading, and without any thought he uses the knowledge and rules which apply to bending strength, even though no parallel exists. Mistakes which have arisen in this way have, in the past, resulted in serious damage. Even if no catastrophes do occur, there are far too many designs which exhibit a built-in weakness to torsional stresses. Perhaps the best-known example of this can be seen in motor-car chassis which are not sufficiently resistant to twisting, and where the torsional stresses are passed on to components neither suited to coping with them nor, indeed, intended to do so (such as the bodywork and open platform members).

It is quite true that for torsional stresses the material should be as far away as possible from the centroid of the section, just as in the case of buckling loads. But unlike buckling and bending, torsional stresses are affected far more by the way in which the individual features making up the cross-section are arranged. This can be explained by offering a simple example: if we take four flat strips and use them to construct (as seen in

Fig. 130.—Junction in a tubular construction, with shaped ends of the tubing tacked ready for welding.

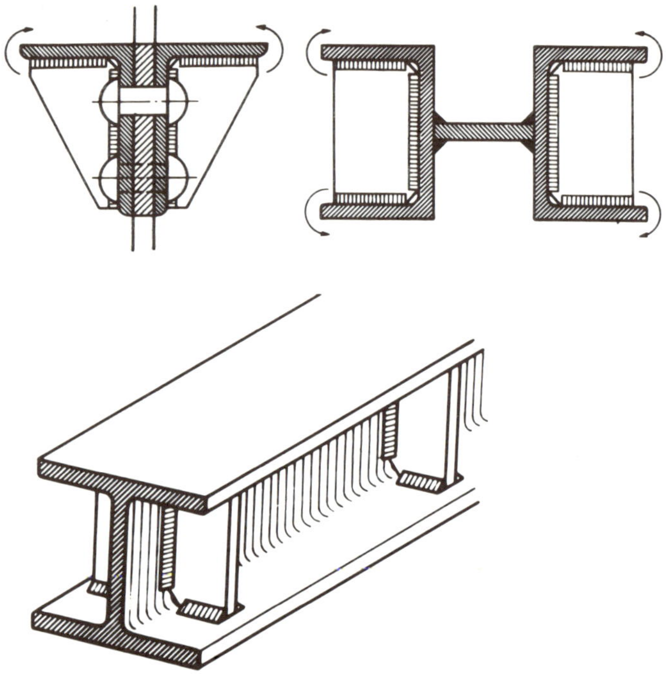

Fig. 131.—Sections used extensively in the past for compression members. The arrows show possible folding over of the wings and flanges, crosswebs being welded in to prevent this happening.

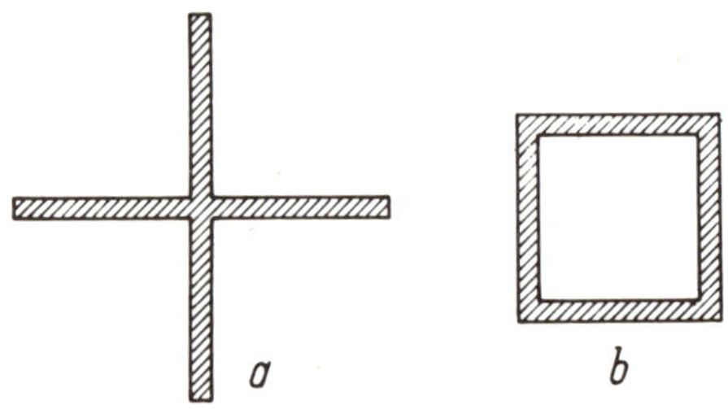

Fig. 132.—Sections built up from identical elements (rectangular flat strips) ; these have more or less the same resistance to bending forces, but very different torsional resistance (1 : 13·7).

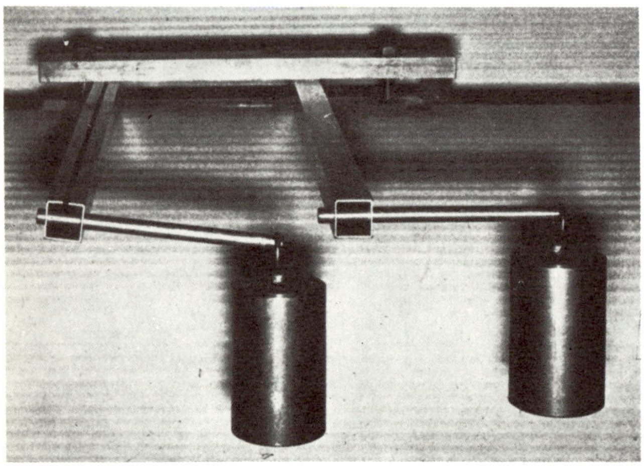

Fig. 133.—Torsion bars made from box-section tubes. On the right, the unmodified hollow section ; left, the same section converted into an open section by a lengthwise cut. The closed section has 80 times the torsional stiffness of that with the cut.

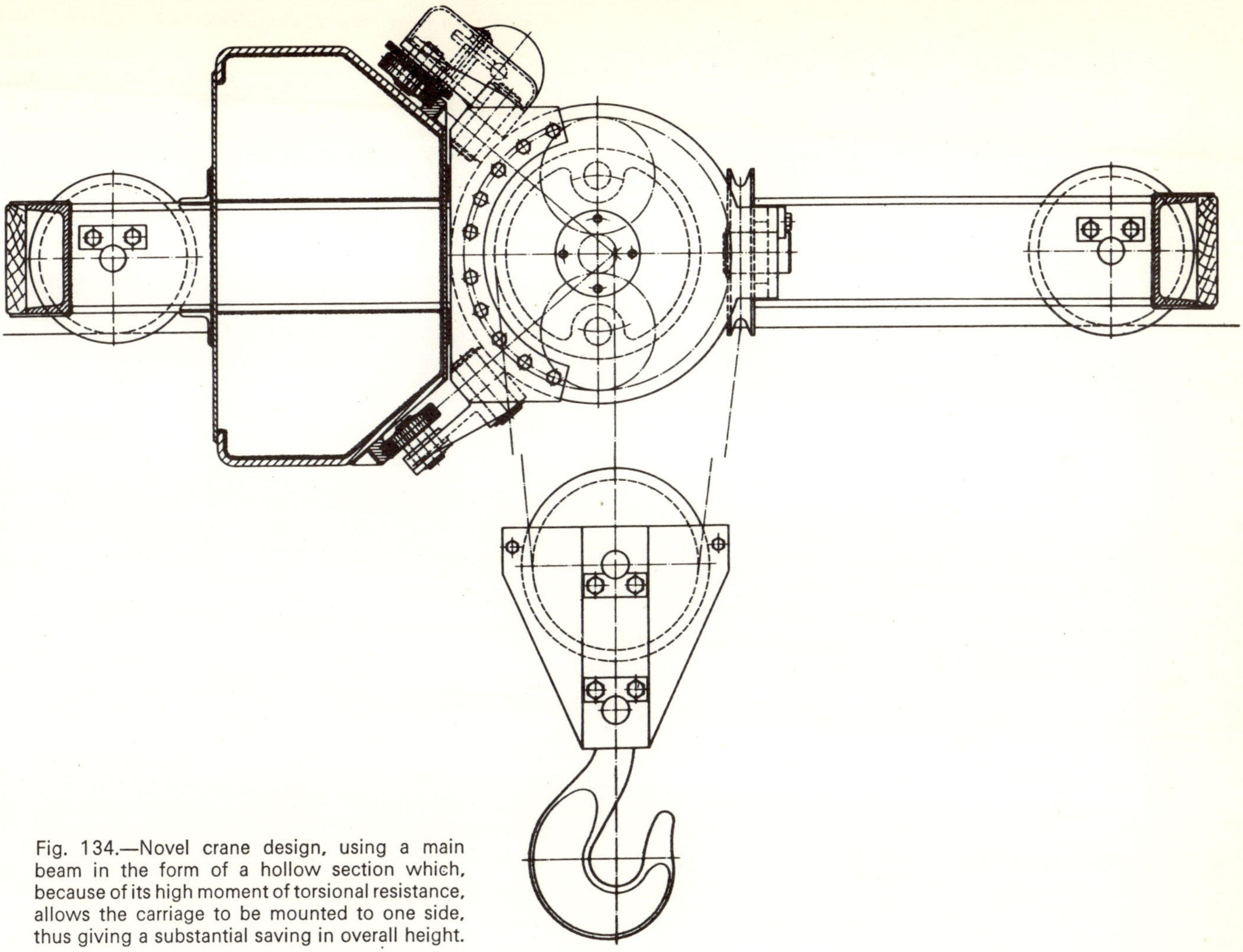

Fig. 134.—Novel crane design, using a main beam in the form of a hollow section which, because of its high moment of torsional resistance, allows the carriage to be mounted to one side, thus giving a substantial saving in overall height.

Fig. 132) in the one case a cross and in the other a square frame, the resistance of sections *a* and *b* to buckling loads will be in a 1 : 1·7 ratio; but their relative resistance to torsional loads will be 1 : 13·7, thus demonstrating that a tubular section is far preferable to an open one (such as our cross). This effect is due to a fundamental difference between shear stress and normal stress, although this is not the place to go into detail. We need only note that closed sections (i.e. hollow beams) are in all cases far better able to withstand torsional loading than open ones, although their second moments of area—axial or polar—may be the same. This is simple to demonstrate by taking a box-section tube and cutting lengthwise down one side to make it into an open section. Both sections will offer more or less the same resistance to stretching and bending, but—as we can see from the experiment shown in Fig. 133—their behaviour when twisted is quite different.

An interesting innovation in crane construction has come

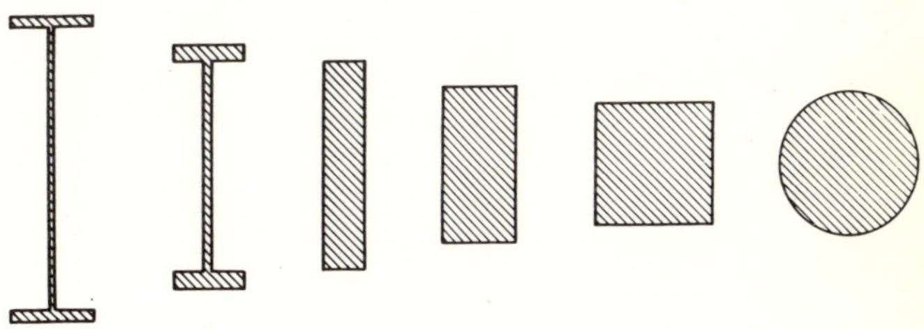

Fig. 135.—Sections of equal bending resistance but widely varying torsional stiffness.

Fig. 136.—Closed sections with equal bending resistance and only slight differences in torsional stiffness.

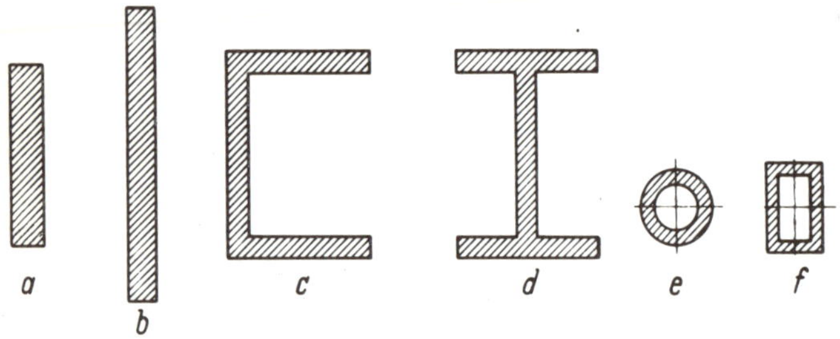

Fig. 137.—Sections of equal torsional resistance. Those at *e* and *f* show the superiority of closed over open sections.

about from the realization that hollow sections are in most instances very much more resistant to twisting than might be supposed. This is the overhead travelling crane using hollow beams, in which the crane carriage runs along the sides of the beams (Fig. 134). Being able to place the carriage alongside the beams at the same height means a substantial reduction in the clearance height of the crane assembly and, since this extends over the whole floor area of the workshop, there is an appreciable reduction in building costs.

Fig. 135 shows a number of sections having the same moment of resistance to bending. Their moments of resistance to torsion are in the relationship 1·0 : 0·63 : 0·37 : 0·17 : 0·02 : 0·01. This drop-off in torsional resistance is not, however, found when we turn to the closed sections seen in Fig. 136; here again, the moment of resistance to bending is the same, but resistance to torsional loads is in a ratio of 1·0 : 0·63 : 1·0 : 0·75 : 0·8 : 0·47. Just how great a difference is made by having a hollow or closed section can be seen from comparing those shown in Fig. 137 *a* to *f*, all of which have the same torsional resistance although their cross-sectional areas vary as 1·0 : 1·26 : 1·55 : 1·55 : 0·32 : 0·41.

Although these facts can be read in textbooks on stresses and strains, one often sees engineering designs which suggest that the designer was ignorant of them. In lorries, for instance, we often find two U-sections facing each other used as a load-bearing frame for the platform. Though they are linked by a number of cross-members, they both act as open sections and thus offer very little resistance to twisting forces. It is obvious that the primary consideration in designing these platforms has been bending stresses, quite overlooking the fact that a lorry is subjected to torsional moments when it is unevenly loaded, or is tipping or is running over rough roads. The answer to this would be to move the load-bearing U-sections together and weld them, or to make each of them into a closed box girder by welding a plate between the jaws of the U.

It was normal in the past to make the bedplates for machinery in a box shape with the open side downwards. Once more, the popularity of this form comes from looking only at one aspect of the problem—regarding the supporting of a piece of

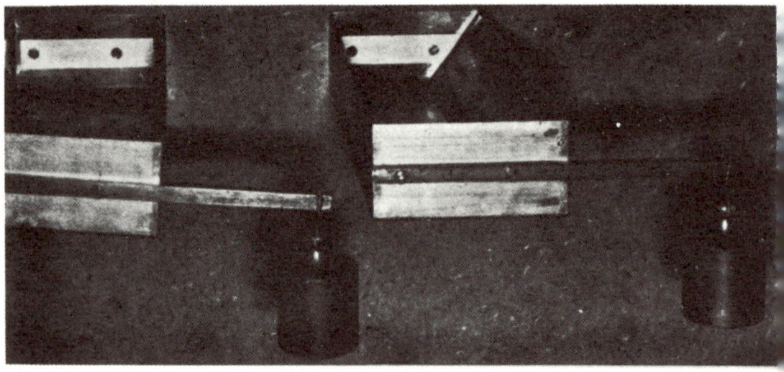

Fig. 138.—Comparison of the torsional resistance of an open box structure (left) and one in a partly closed form (right).

Fig. 139.—Test-brake stand using tubes as load-bearing members.

machinery as solely a flexing problem and forgetting that the bedplate must resist twisting as well. It will be obvious from what we have just been discussing that a box-shaped frame will not be able to do this. Fig. 138 shows, on the left, an open section of this kind which is clearly unable to offer any appreciable resistance to the twisting force exerted by the lever and weight. The construction on the right is exactly similar except for a strip of metal which has been riveted at an angle to the right-hand side of the box. The ⊔ section has been made into a partly closed ⊿ section, with the result that the box is able to cope with the twisting load with practically no distortion at all.

This principle ought to be given far more attention everywhere in the designing of bedplates and supporting frames for machinery. In Figs. 139 and 140 we see two bedplates in which the load-bearing components are a pair of tubes; they are generally far more resistant to torsional stresses (while at the same time being much lighter) than any other form of bedplate.

The experiment we have just seen can also offer valuable lessons for the stiffening of flat walls. As a rule, attempts to stiffen these take the form of added ribs. These are certainly effective against bending stresses, but not always so against torsional loads; and then people are surprised when things like doors twist alarmingly under a slight twisting moment even though stiffening and raised edges have been provided in an attempt to prevent this happening (Fig. 141).

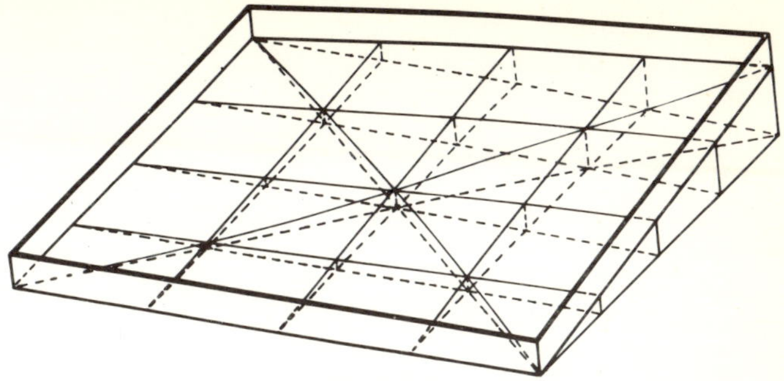

Fig. 142.—Flat sheet, with little resistance to twisting. The orthogonal network of lines remains undistorted, the right-angles unchanged, but the diagonals are bent.

Fig. 140.—Bedframe for a convertor unit made from tubing. This frame is not only lighter than the orthodox open-box structure, but substantially stiffer as well.

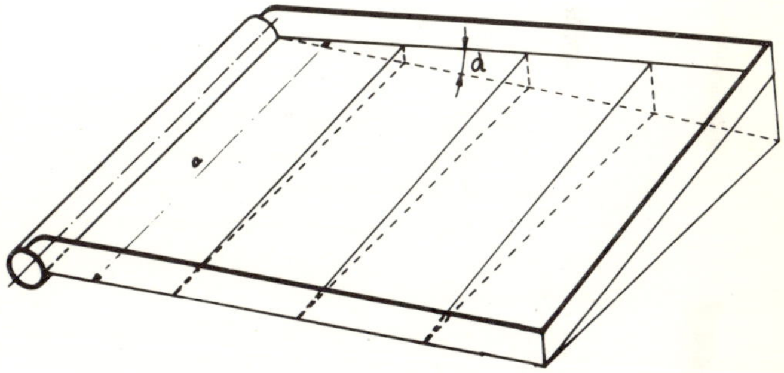

Fig. 143.—Adding a tube as part of the edge of the sheet is enough to stop it twisting.

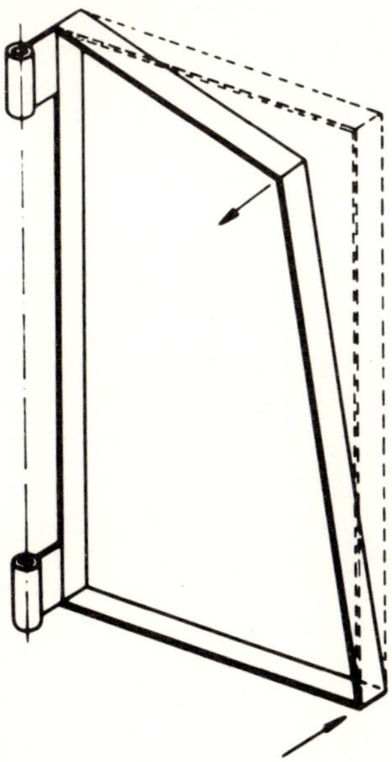

Fig. 141.—Doors stiffened by means of raised edges can still be easily twisted askew.

It is easy to imagine how a flat plate can be twisted without the edge being bent. The plate itself bends, one diagonal being twisted one way and the second diagonal the other (Fig. 142). If the plate is thin, it offers no resistance worth talking about to this distortion, nor do the edges, which undergo only one twist. The only way of bringing about a change in this behaviour is by adding a rib to stiffen one of the diagonals, or by making one of the four edges into a hollow section (Fig. 143). A tube of this kind would have to allow itself to be twisted through the angle α along the length a across the plate if the plate were distorted in the way shown in the diagram. Because of its substantial second moment of area, however, the tube would strongly resist such a distortion and would therefore protect the plate against twisting.

We might mention here one basic shortcoming of thin walls as a constructional element—their flexibility. A large sheet of metal can be pushed in appreciably by hand pressure, even though it may be quite thick. This is why we often see waves and ripples in parts made from thin sheet, which appear at the slightest stress and which even skilled treatment will fail to remove. A familiar example of this is illustrated in Fig. 144. Uneven panels like the one shown not only look ugly, but also have the functional disadvantage in that they tend to flap and oscillate—they act, in fact, like membranes, responding to the slightest disturbance and making a drumming noise. The possibility of this occurring must be carefully considered in gearboxes, where the flat walls of a housing may act as a sounding board. If the ratio of wall area to thickness becomes excessive, the wall will need stiffening with ribs (Figs. 145 and 146). The most important thing here is the pattern formed by the ribs as a whole. Ribs running in only one direction, i.e. a strip pattern (Fig. 147*a*), as well as any orthogonal pattern (Fig. 147*b*), will be resistant to bending forces, but not to twisting. A triangular pattern, on the other hand, will resist twisting as well (Fig. 147*c*). This is because jointed quadrangles can be distorted to form parallelograms, while a jointed triangle will always keep its shape; and in these circumstances such rib systems can be likened to jointed systems. The difference in resistance to twisting can amount to several powers of ten, and is especially obvious from Fig. 148. This should be a convincing reason for using only triangular ribbing for mechanical systems (such as machine tools) which need to be particularly rigid.

The importance of torsional resistance in a structure is generally underestimated, and one dramatic case where the results attracted world-wide attention was the collapse of the

Fig. 145.—Thin-walled shell for a marine gearbox. Wide welded-on ribs give the stiffness needed for rigid positioning of the gears. (*Photo: Brown Boveri, Baden.*)

Fig. 144.—Thin panels can never be made perfectly flat. The play of light on the outside of this tram bodywork betrays a considerable lack of flatness.

Fig. 146.—Thin-walled exhaust connector for a large steam turbine, stiffened with a close network of orthogonal ribs (Brown Boveri, Baden).

Tacoma Bridge in the United States. A few months after its completion and opening to traffic, this suspension bridge was set in torsional oscillation by a strong sidewind (Fig. 149*a*); the oscillation reached such proportions that the bridge was unable to withstand the stresses produced, and fell apart (Fig. 149*b*). The building up of the oscillation took some time, and there was plenty of opportunity to photograph the moment of disaster.

It is quite evident from the behaviour of the bridge that the torsional stiffness was inadequate. The designer of the bridge cannot be accused of neglecting the twisting problem, however, for the stiffening ribs that can be seen on the underside of the load-bearing surface are quite clearly in a triangular pattern, which is far more resistant to twisting than a rectangular one. The fact that such an accident could nevertheless occur, is proof that in spite of the suitable layout of the ribs the degree of stiffness was still not sufficient; clearly this was because the ribs were not thick enough, and probably not deep enough either. This impression is borne out by the glimpse we get of the underside of the falling carriageway in Fig. 149*b*. The design was subsequently amended on aerodynamic grounds, by adding slots along the side walls; whether anything was changed in the mechanical design of the bridge, I do not know.

If the ribs take the form of a hollow section, one is not limited to a particular pattern; this is an advantage from the outset, since triangular systems are seldom attractive in

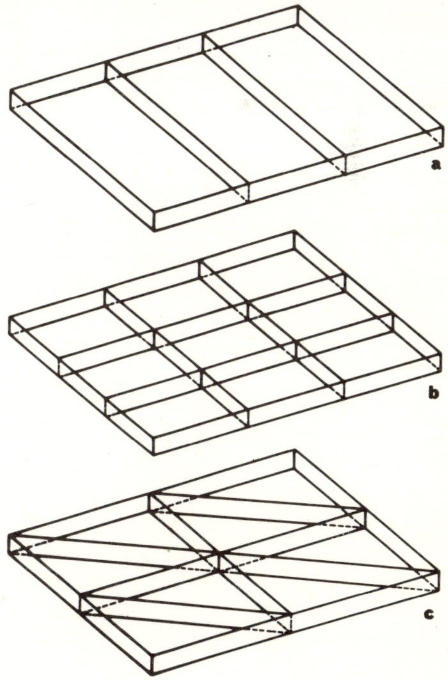

Fig. 147.—Patterns for welded-on ribbing. *a* and *b* have low, and *c* high, resistance to twisting forces.

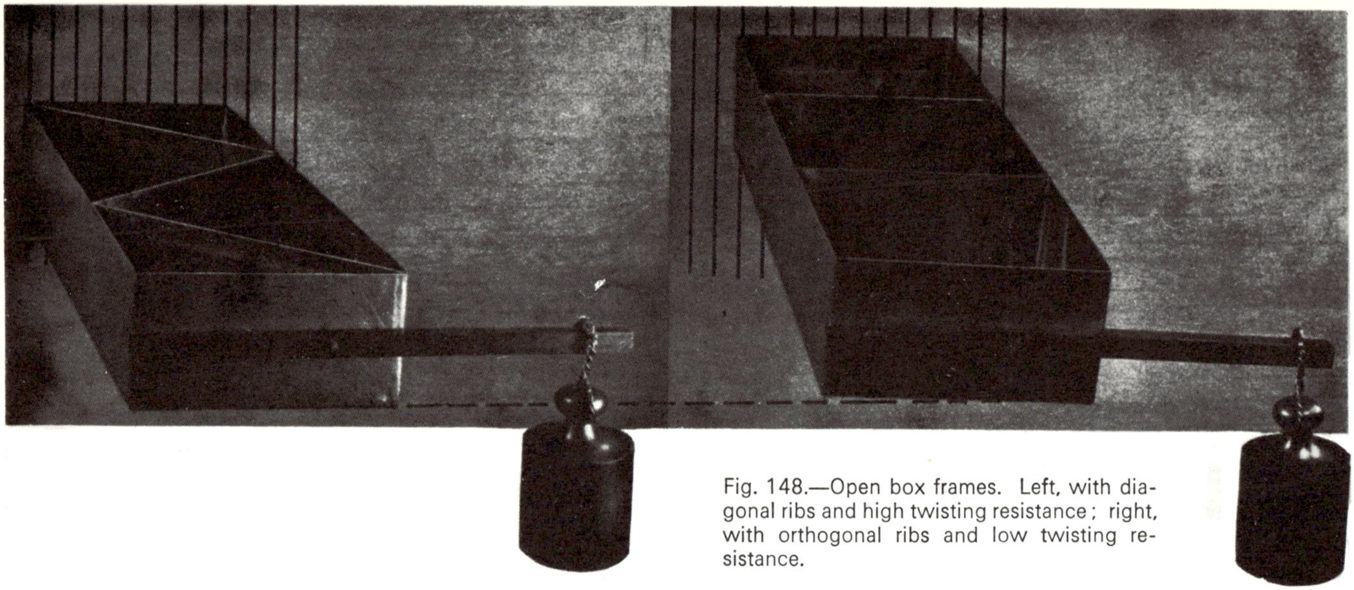

Fig. 148.—Open box frames. Left, with diagonal ribs and high twisting resistance; right, with orthogonal ribs and low twisting resistance.

appearance. From the point of view of casting technique, however, producing hollow ribs presents a great deal of trouble, for any hollow space needs to be filled with a core during moulding. Consequently it is hardly possible to have hollow sections in a casting, although it is quite simple to have hollow ribs in a welded construction. Fig. 150 shows a number of ways of applying standard sections to a sheet, thus producing hollow ribs. This method is already widely used in lorries and builders' plant (Figs. 151*a* and *b*). Where it is impossible, for production or functional reasons, to make use of ribs for stiffening flat areas, then the completely flat form must be abandoned and curved shapes used instead. Conical and cylindrical shapes are the easiest to produce, although even here stiffening may sometimes be needed if the areas involved are very large (Fig. 152). The housing seen in Fig. 153 is of interest in this connection; it was made for the low-pressure section of what was, at the time, the largest steam turbine in the world (160,000 kW). The ribs fade out at the point where the single-curvature surface of the shell takes on a compound curvature.

Where the aesthetic appearance is important, the shell type of design is indisputably superior to a ribbed construction, even though it may be very much dearer to make. It is, incidentally, possible to provide a kind of false ribbing by stamping the sheet with corrugations (a process known as 'swaging'); the appearance is not unattractive, and some years ago the method became particularly popular in motor vehicle bodywork. With all these stiffening elements we must be quite clear in our minds that they alone will have to take any bending or twisting stresses that occur, as they get virtually no assistance from the flat surface to which they are applied.

It might be useful, speaking of parts taking a bending stress, to give one or two general comments on suitable and less suitable shapes. The material is being used to best advantage—from the viewpoint of bending loads—when it is away from the centroidal axis (in the direction of the bending plane, of course, and not crosswise to it). For lightness of construction, therefore, the choice will be tall I-sections with a thin web and thick flanges. A thin web can, however, buckle; so either it must not be made too thin, or it should be given support at the sides to prevent buckling (Fig. 154). Where a saving in material is especially important, lightening holes may be provided in the web. This is particularly common in motor vehicle construction. Care must be taken, however, that the web does not become merely a lightly-loaded link between the flanges carrying the main load—it plays a major part in carrying the shear forces. It should be noted that shear stresses reach a maximum in the web—at the centroidal axis, in fact—and not in the flanges. If lightening holes are made in the web, the flow of the shear forces will be interrupted at the very point where they reach their maximum; as a result, there is a risk of the web shearing between the lightening holes (Fig. 155). Where flexing occurs without axial tension or compression, it is quite correct to choose sections which are symmetrical with respect to the bending axis, i.e. rectangular, I or box sections. But as soon as there is tension as well as flexion, the side of the beam that is under tension suffers more than the side under compression, and more material should be moved to the

Fig. 149 *a* and *b*.—Suspension bridge at Tacoma (Washington, U.S.A.). This bridge, with a span of 850 m (2800 ft) collapsed in 1940 four months after it was opened. Side winds set up torsional oscillations in the carriageway, which obviously had inadequate torsional stiffening and finally gave way under the stresses and fell apart.

former side. This means that in such instances an asymmetrical section is better than a symmetrical one, and the choice will fall on a T-section or something similar. The same applies to materials which do not have the same resistance to tensile and compressive stresses; an example of this is cast iron, which has some three to five times more resistance to compression than it has to tension. So with a cast iron component, too, the 'tensile' side must be reinforced with extra material, again resulting in an asymmetrical section.

Special conditions also obtain in the case of parts carrying a bending stress which have a pronounced curved axis (such as the crane hook seen in Fig. 134). In a support of this type the stress distribution across the section does not follow a linear

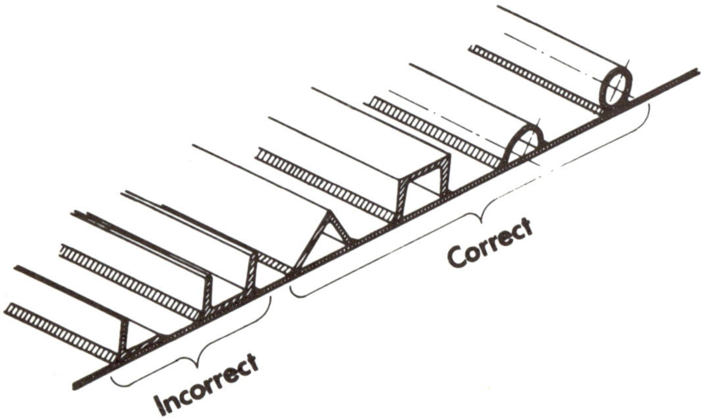

Fig. 150.—Closed sections, which will be resistant to twisting as well as bending loads, can be produced by attaching open sections to a flat panel.

Fig. 151*b*.—Tipping hopper truck. The floor of the hopper has been stiffened by swaging, but because these corrugations are open underneath there is little torsional resistance. The bridge of the tipping gear has, however, been fabricated from hollow sections (E. Wirz, Uetikon ZH).

Fig. 151*a*.—Side walls of a dumper truck stiffened with deep ribs. If the ribs are made as hollow sections, this will also give high torsional stiffness (E. Wirz, Uetikon ZH).

law as in the case of a straight bar or rod, but a hyperbola, one asymptote of which is perpendicular to the centrepoint of curvature on the cross-sectional area concerned. As a result there is usually a steep increase in stress along the inner edge of a sharply curved beam, which may easily be several times the maximum stress on the outer edge. Once again, the inner edge will need to be strengthened, and the section will be asymmetrical.

What is frequently overlooked is that any bend in an initially straight beam will make it behave in the same way as a curved beam (Fig. 156), with the same typical stress distribution. So any such changes of direction must not be too abrupt, but must follow a long gentle curve; and whenever possible some sort of reinforcement should be added on the inside of the bend (Fig. 156*b*). What should not be done (though previous unfortunate experiences might prompt it) is to weld in a knee plate (as seen in Fig. 156*c*) to overcome the disadvantage of a small radius of curvature. If an unstrengthened beam like that in Fig. 156*a* tends to crack at point *x*, one 'reinforced' as shown in Fig. 156*c* will be even more likely to do so at point *y*.

Fig. 152.—Simple curved panels. A tangential pattern of ribs is enough to give the strength needed (Brown Boveri, Baden).

Fig. 153.—An example of how a shell with a single curvature may distort in one direction and must be stiffened with ribs, but not when there is a compound curvature. Note how the ribs are discontinued when the double-curved area is reached. (*Photo: Brown Boveri, Baden.*)

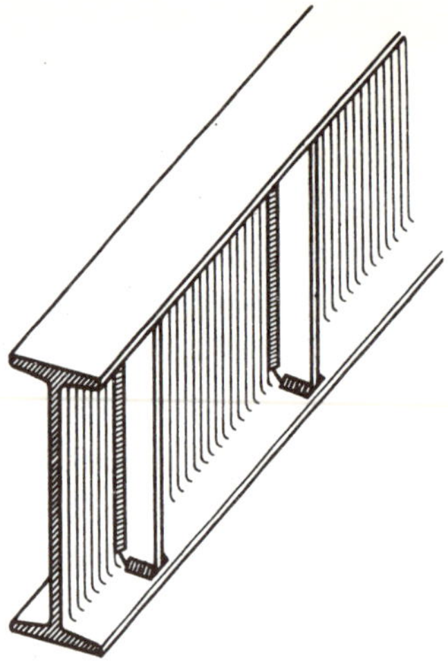

Fig. 154.—The web of a tall section under a bending load may buckle or bulge, so crosswebs are added at regular intervals.

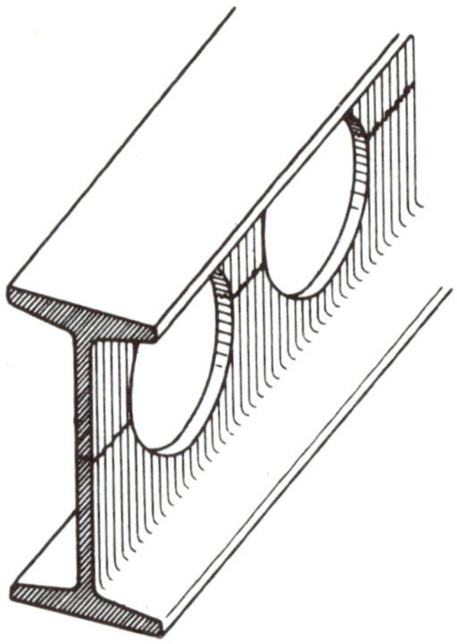

Fig. 155.—Over-large lightening holes in the web of a tall section under a bending load may lead to shearing through the remainder of the web, where there is a concentration of shear stresses (note line of fracture along the longitudinal axis).

The knee plate lies in the direct line of force and carries the main load, and is thus particularly vulnerable. If knees are added in this way, fractures starting from the inside of the bend are almost inevitable.

The phenomenon of 'Lines of Force'

It must be said at once that the expression 'lines of force' or 'strength flow' which is used so often in design work does not have a scientifically-based definition like that of state of stress or second moment of area. It does, however, provide a convenient means of visualizing and explaining certain situations. The concept comes about from the observed fact that when a body is subjected to a mechanical load, the forces exerted on one part are passed on to the next so that there is a flow of these forces which, like a river, follows a given path through the various parts of the machine. If we examine the way the load is supported in a thrust bearing like that seen in Fig. 157, it is easy to trace the path followed by the lines of force through the various parts. They appear at the cross-section of the shaft at *1*, pass up the shaft to the head at *2*, are transferred to the supporting ring *3* where they divide to reach the bearing surface via two paths, come together again and continue in a broad flow (which is not evenly distributed over the width and circumference) into the pressure segments at *4*, merge and pass into the pressure seating *5* and thence away to the support at *6*. It is not possible to get a precise picture of the individual stages in this sequence, but it is quite certain that the lines of force follow definite and chosen paths and avoid others, and

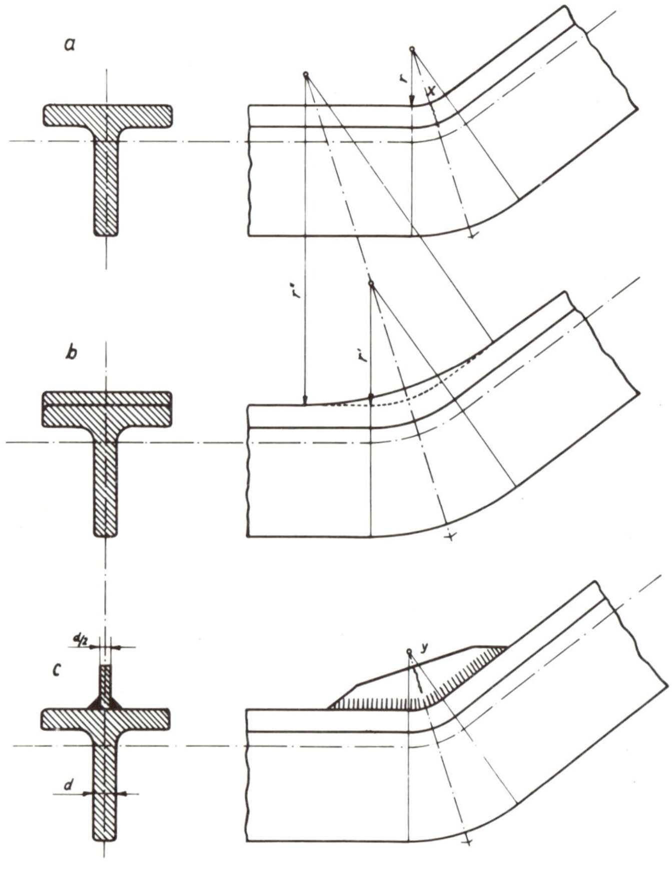

Fig. 156.—Initially straight bars which are bent act as sharply curved beams, and have a different stress pattern, with stress concentrations at the inside of the bend and cracks appearing at *x* (see *a*). Reinforcement, as in *b*, is therefore needed. It is wrong to add knees on the inside, as fractures at *y* are then particularly likely.

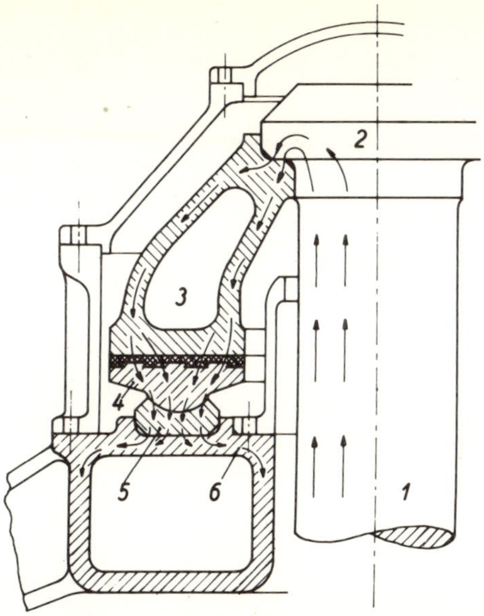

Fig. 157.—Pattern of lines of force through the various parts of an axial bearing.

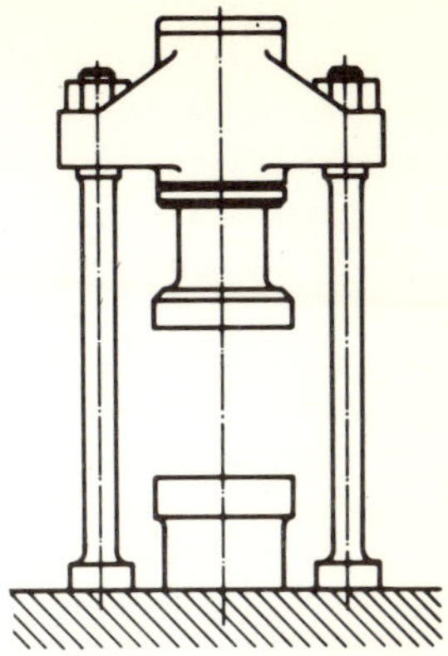

Fig. 158.—Diagrammatic sketch of a press, an example of how the strength flow in various parts of a machine can have varying values, so that the crossbeam needs to have several times the cross-sectional area of the two pillars.

that in one place they flow close together while in another they will be spread apart.

We can conclude from this that at various places in a machine component there will be a widely varying density of strength flow (in other words, a varying degree of stress on the material), and that it is imperative for the designer, in deciding on shapes and dimensions, to make allowance for this distribution and especially for the main concentrations of lines of force. He does this by means of stress studies, which usually take the form of calculations and may also involve tests on models. This represents a not inconsiderable part of the designer's programme of work, and he may often need specialist assistance. Stress studies on something like a spiral casing or a multi-throw crankshaft may take days or even weeks to complete. In most cases it is considered enough to determine the stress distribution at a few specially vulnerable points; but there are instances where it is thought necessary to present the whole three-dimensional stress distribution in a single equation and thus determine the complete pattern of the strength flow.

Very often the lines of force travel along distances between two widely-separated parts of the design. As an example of this we may consider the internal-combustion engine. We can see, in Fig. 160, how the lines of force which arise through the pressure of the gas on the piston crown pass through the gudgeon pin, the piston rod, crosshead (not present in Fig. 160) and connecting rod to the crankpin, whence they run through all the adjoining crankpins and crank webs until they reach the clutch. In their travels, these lines of force change their nature several times, appearing sometimes as direct compressive stress and sometimes as bending stress, and finally taking the form of a torque. Since without a pressure on the piston crown there would be no torque at the clutch plate, it is obvious that we are dealing throughout with different facets of one and the same phenomenon.

It can happen, too, that the strength flow is split up along the way. There will not, for instance, be a pressure on the crankpin without a matching load on the crankshaft main bearings. These are seated in the crankcase cross members, to which the pressure is transferred and from which it passes, usually via tie rods, into the cylinder block and then through a number of studs into the cylinder head; the exploding charge of gas exerts a thrust against the cylinder head in order to press down on the piston crown. We notice that this secondary path for the lines of force forms a closed circuit—a peculiarity of strength flow which we often encounter.

It is not always easy to keep this phenomenon in the mind's eye in designing when it takes so many different forms, and it is not surprising that the designer occasionally loses sight of it and forgets to check the load capacity of one cross-section or another.

To prevent this happening it is useful, with any design that involves mechanical stresses, to begin by tracing out the overall pattern of the lines of forces and only then to go on to working out the detailed treatment of the design.

Effect of shape on loading and stresses

Having said in the previous chapter that the strength flow occasionally changes its nature along its path, we must add that the force intensity involved may also vary considerably. The longitudinal tensions in the upright columns of a press frame (Fig. 158) give rise to compressive and tensile stresses in the yoke at the top which may be many times that of the tensile stress in the columns. Anyone dealing regularly with strength of materials problems will be well aware of these relationships, and even in extreme cases they should not come as a surprise. It is important here, however, to note that this increase and decrease in the density of the lines of force is very largely dependent upon the shape of the parts in question, and that the designer himself is in a position to decide how large the variations in strength flow are going to be. Let us compare three ways of transmitting forces, seen in Fig. 159: a straight bar as at I, a kind of chain link II, and a crescent-shaped beam III. All of them carry the same maximum stress, so that the extravagant use of materials in III is striking; it can only be due to an increase in the strength flow in the curved beam.

The lesson to be learned from this is of fundamental importance: it is that the lines of forces should always be passed from one point to the next by the most direct route possible, which means that the central axes of mechanically-stressed parts should follow a gently flowing line.

Put in such ultra-simple terms, this statement may seem almost banal; and yet experience shows that it can be very useful to keep this point in mind. At all events, it took designers scores of years to realize its importance and to learn to apply the principle consistently in their machinery. If, for instance, we look at the Diesel engine in Fig. 160*a*—a design built some 50 years or so ago—we are at once struck by the wide spacing of the two cylinder supports, and we can see that there must be considerable bending loads on the crosswise wall of the crank-

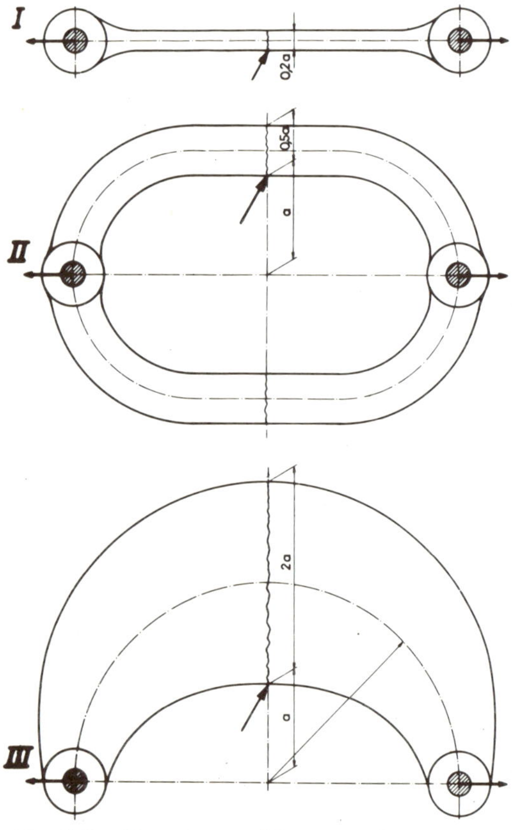

Fig. 159.—Three types of force transference compared; in each case the maximum stress at the point arrowed is the same.

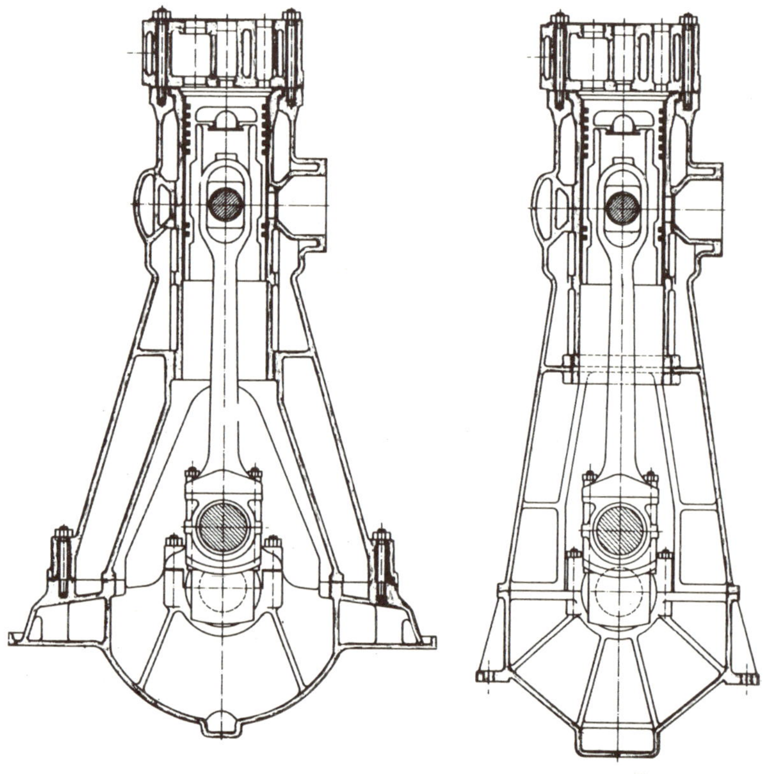

Fig. 160*a*.—Old design of Diesel engine seen in cross-section as an example of poor strength flow pattern in the frame.

Fig. 160*b*.—Better design for the same engine, with the vertical supports brought inwards, close to the crankshaft main bearings.

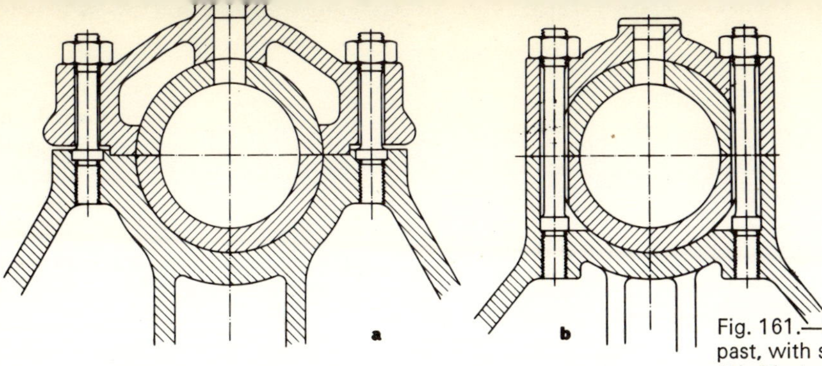

Fig. 161.—(*a*) Bearing design common in the past, with strength flow extended sideways. (*b*) Modern design, with bearing cap bolts close to the shaft, giving a shorter path for the lines of force and a higher load capacity.

case as a result. These loads could be reduced, and the whole engine frame made much lighter, by bringing the cylinder supports inwards, closer to the crankshaft main bearings (as in Fig. 160*b*).

Further examples from practical engineering are shown in Figs. 161*a* and 162*a*, where the lines of force have been led along unnecessarily long paths, and needless extravagance in material has resulted. Figs. 161*b* and 162*b* show how the shapes can be improved and the strength flow given a more favourable pattern. Many more examples could be given to prove the importance of these relationships.

Figs. 163 and 164 demonstrate that, alongside less successful designs, there are others with an excellent lines-of-force pattern; both of these have, however, been taken from modern engineering practice.

A further lesson to be learnt from Fig. 159 is that a symmetrical arrangement is better than an asymmetrical one. The

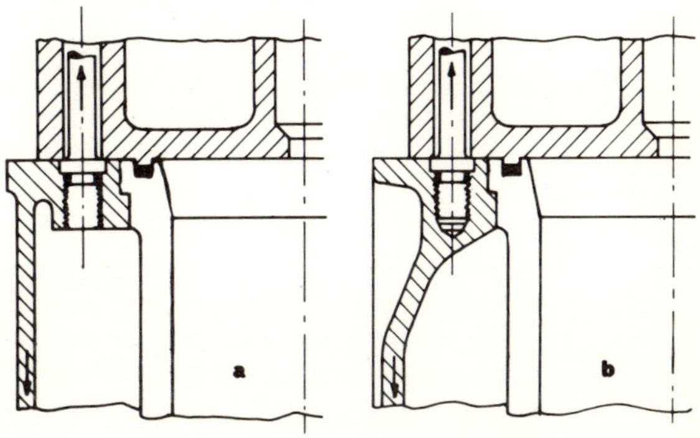

Fig. 162.—(*a*) Poor strength flow pattern through cylinder head stud, with the lines of force running upwards from the stud before being led down the wall.
(*b*) Improved flow pattern, with the lines of force running directly downwards.

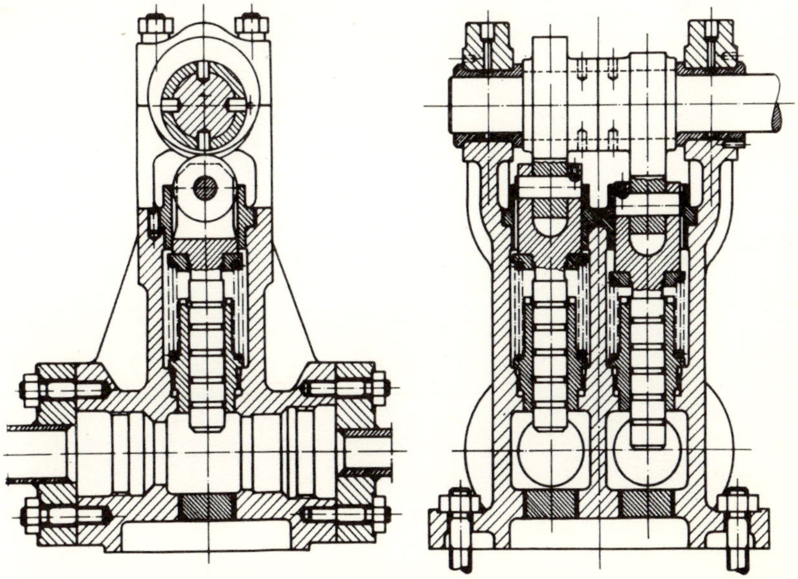

Fig. 163.—Small twin-piston pump, a good example of bearings located close together giving a short path for the strength flow.

tension members at II and III are, as we have said, carrying the same load; yet it is obvious at a glance that the curved beam is nearly twice as heavy as the chain link. For this reason a lifting eye will always be better than a lifting hook (Fig. 165). The riveter stand in Fig. 166 illustrates the extreme limits to which strength flow can increase if led through very circuitous paths. It seems almost incredible that the tiny press at the left-hand end of the arm can stress the massive bow-shaped beam on the right to its full capacity.

It is all the more reasonable to lead the lines of force along the shortest path, since this in fact is the route they themselves try to take between one stress-point and the next. If we give the lines of force plenty of room to find their own path as in Fig. 167*a* they will always choose the shortest way between stress-points and concentrate in the area between these. Even though the strength flow may not follow a thin line through the plate, the greater part will nevertheless be concentrated in the centre of it and there will be a sharp drop-off towards the edges (diagram to right of Fig. 167*a*). The lines of force invariably try to take this shortest and simple path through a part, even if the external contours of the part make this difficult. If we cut a crescent-shaped piece out of one side of the plate and (to keep the cross-sectional area the same) replace it on the opposite side as seen in Fig. 167*b*, the lines of force will pass through the centre of the new cross-section; yet to maintain the equilibrium they will take up a distribution such that they again concentrate along the straight line joining the points of stress. The distribution is shown on the right of Fig. 167*b*; as can be seen it results in an extreme concentration along the edge of the area where the part has been removed. The peak stress arising here may easily be several times the previous

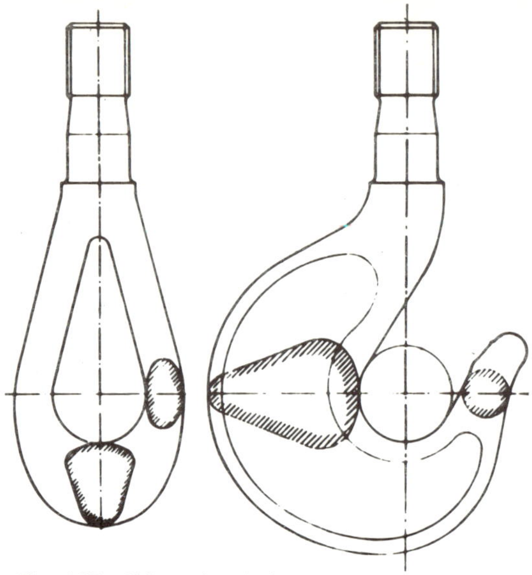

Fig. 165.—When the designer can choose between a symmetrical or asymmetrical strength flow, the first is always preferable. A lifting eye has many advantages over a lifting hook.

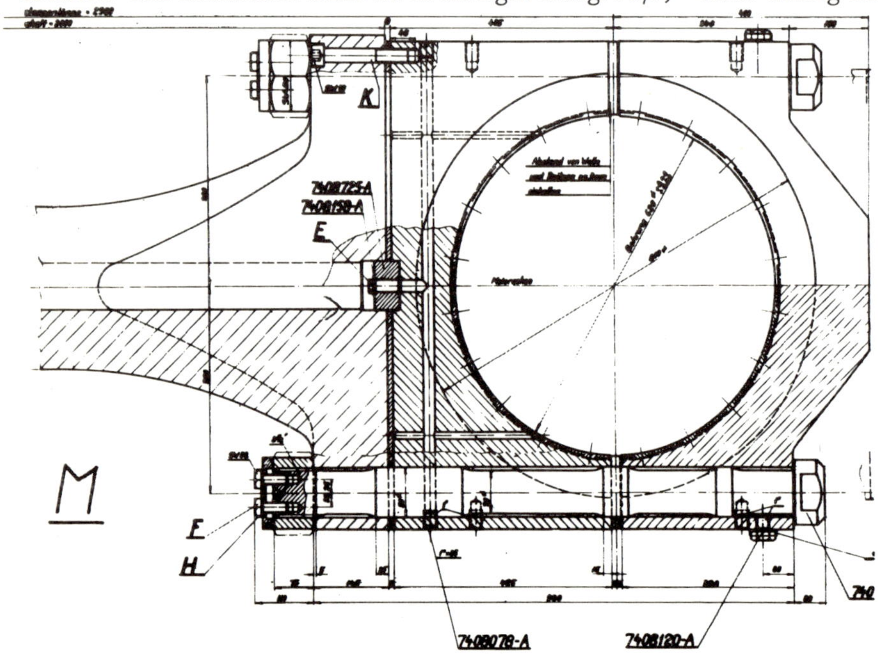

Fig. 164.—Push-rod head of a large Diesel engine (900 mm bore), with the bolts extremely close to the shaft—an ideal way of saving weight (Sulzer Bros., Winterthur).

maximum. If the cut-out section is made even deeper (Fig. 167*c*), this effect is magnified to the extent that the stress on the opposite side not only falls to zero but may even become compressive (see diagram on right in Fig. 167*c*).

In a similar way the lines of force may be deflected from their normal path by irregularities inside a workpiece, e.g. by a hole or a transverse fissure (Fig. 168). In both cases there is an increase in stress where the edge of the irregularity cuts into the path of the lines of force, sharp edges giving rise to particularly high peaks. It will be obvious that such stress peaks will detract from the strength of a machine part.

Fig. 166.—Riveter stand, with enormous increase of stress in the bow caused by high loading at the jaws carrying the riveting press.

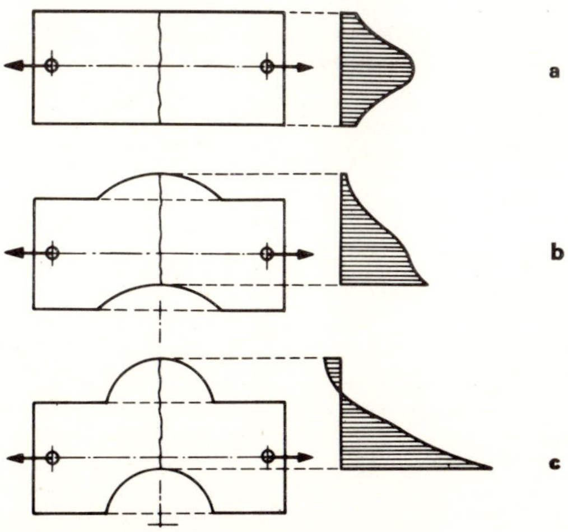

Fig. 167 *a, b, c*.—The profile of the workpiece has a decisive effect on the distribution of strength flow within it.

If cutting into the low-load marginal areas of the strength flow results in disturbance to its pattern and increases in stress, this will be even more the case when the lines of force are already densely concentrated at the point in question. This is the case with bending or torsional loads, where the stresses are distributed over the cross-sectional area in such a way that they reach their maximum values along the outside edges (Fig. 169*a*). If the profile cuts into the lines of force at a spot like this, the stress peak that results is particularly high (Fig. 169*b*) and there may easily be a fracture from a fairly modest load.

We are already aware of this effect from everyday experience—we know, for instance, that it is far easier to break a rod in two if we cut a notch at the point where we want the break to occur—the same applies to cutting glass. Glass is, in fact, not really 'cut' at all; it is broken after a fine notch has been scratched in its surface at the desired place with a diamond 'glass-cutting tool'. The concept of the notch has been carried from observations like these into modern strength of materials theory, and has become a new and wide field of research concerned with the effects of grooves and notches on the strength of machine parts. The results of this research are incorporated in what is known as *notch stress theory*, and they have one characteristic in common—the more directly and sharply a notch cuts into the area of the lines of force, the

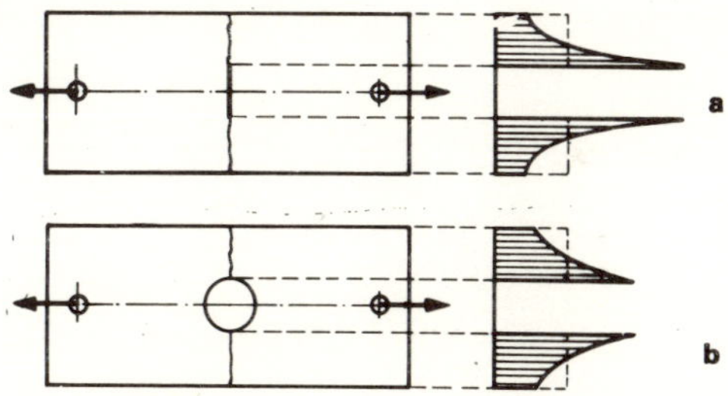

Fig. 168 *a* and *b*.—Discontinuities inside a workpiece stress peak, varying with the shape of the irregularity.

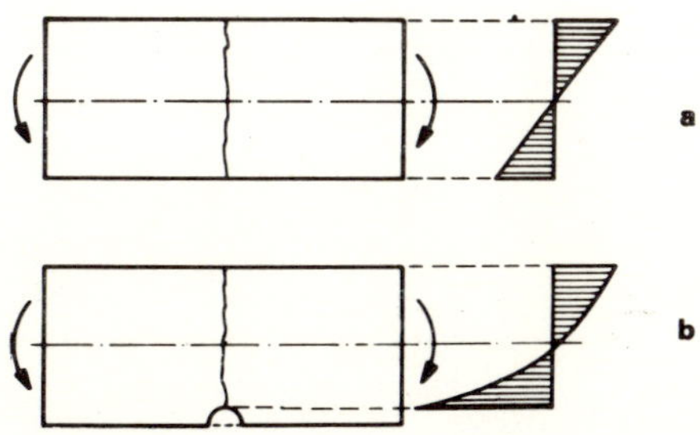

Fig. 169 *a* and *b*.—Irregular outlines produce stress peaks especially in highly-stressed peripheral areas.

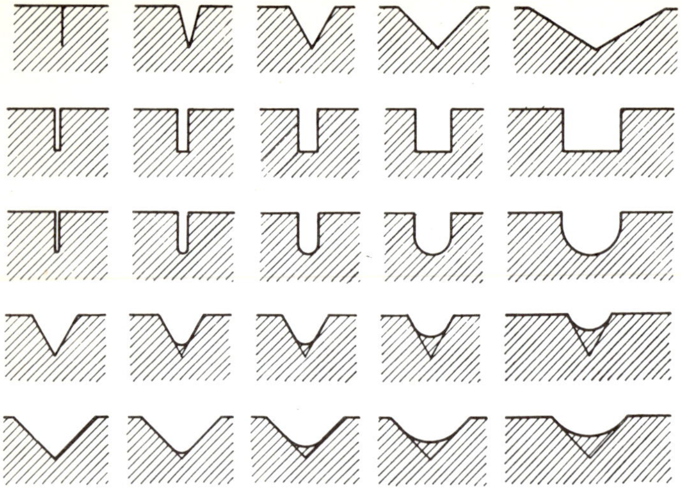

Fig. 170.—Common notch shapes, arranged in order of their effect on the strength flow (decreasing from left to right and top to bottom).

Fig. 171.—Asymmetric notch shapes, arranged in the same way as in Fig. 170.

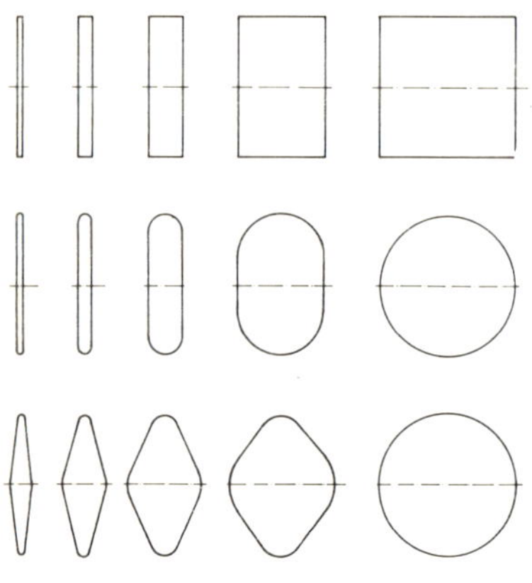

Fig. 172.—Internal notch shapes, arranged in the same way as in Fig. 170.

greater the stress will be at that point. Because of this, the various kinds of notch can be classified according to the extent to which they upset the flow of the lines of force, as shown in Fig. 170.

Notches may also be asymmetrical, or occur inside a workpiece. There is an enormous range of possibilities, and Figs. 171 and 172 illustrate a few of these. Here again, they are arranged with the notch effect becoming less from the top downwards and from left to right.

It is important, from the designer's viewpoint, to be able to calculate accurately the effect of a notch on the stress pattern. One of the main preoccupations of research has therefore been to provide a basis for calculation; it will be obvious—particularly in view of the many kinds of notch shapes possible—that the range of data for calculations is extremely wide. It would be going beyond our scope to give even the most important of these data; such a survey is, however, contained in the Swiss VSM Standard Sheet 14331, which the student is urged to study. There is a similar survey in English, ESDU 65004 *Stress Concentration Data* published by the Engineering Sciences Data Unit.

We have already pointed out that the effect of a notch on the stress pattern is determined not only by its shape but also by the kind of stress involved, and the general pattern of the lines of force at the point in question. It matters a great deal, for instance, whether the notch is located so that it merely transfers a straight-line strength flow (Fig. 173*a*) or whether it also reverses the direction of that flow (as in Fig. 173*b*). In the latter case two effects occur together and are cumulative, so that the density of the lines of force is particularly high at transition points like these. We shall see later that this state of affairs

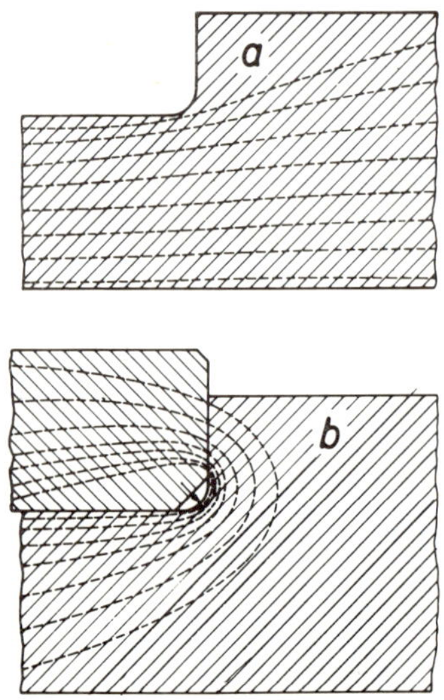

Fig. 173.—Two parts with the same shape (step in cross-section) but differing strength flow pattern can give totally different notch effects, and widely differing stress levels at the corner of the step.

applies to clamping bolts, and is one of the reasons why these bolts frequently snap. The fact that the effect of notches will differ with various kinds of strength flow is seen especially clearly when one perforated bar is subjected to a tensile load and another to a bending load. The stress pattern under tension is shown in Fig. 168*b*, and that under bending in Fig. 174, with the calculated stress pattern in an unperforated bar shown as a dotted line in each case. The effect of the hole on stress distribution under flexure is quite different from that under tensile load and is, as can be seen at once, largely harmless. The hole produces a local increase in stress where the stress level is initially much lower than at the outer edges and thus quite able to cope with an increase. This increase in stress makes such a substantial contribution towards the amount of resistance that it even results in a reduction of the stress at the outer edges; and so it comes about that a hole in the centreline of a bar bending has no detrimental effect and may even lead to a reduction in the peak stresses in the outermost layers.

This is, however, only true for direct stress situations, and in the case of shear stresses the very opposite applies. In a bar under a bending load the shear stresses reach their maximum at the centroidal axis (Fig. 175), and if a hole is drilled here extremely high stress peaks may occur. These do not, it is true, appear at the edges of the hole, since from the law of equality of complementary shear stresses it is impossible for shear stresses to appear at a free surface. This does not, however, prevent there being a sharp rise in stress here, so that a maximum occurs very close to it. If the shear stress is large in relation to the bending stress, the stress peak may be very dangerous.

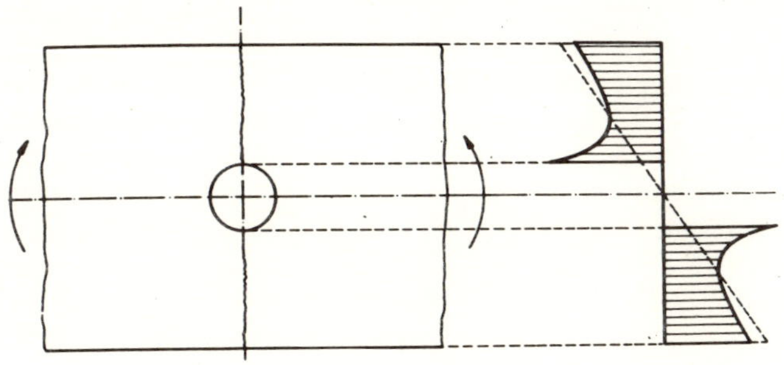

Fig. 174.—A hole in a bar under bending produces only a small change in the normal stress pattern.

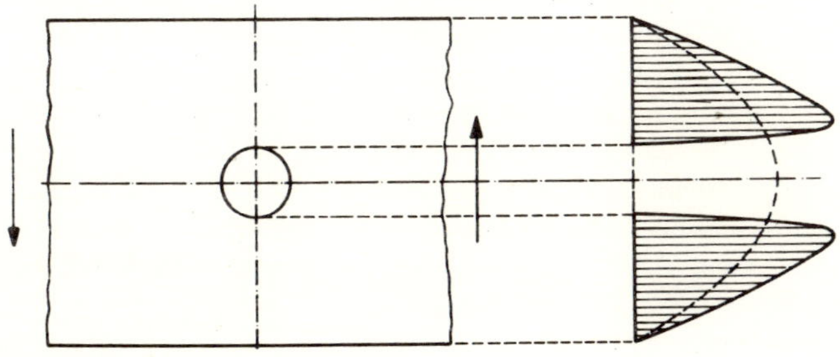

Fig. 175.—The same hole has a much more serious effect on the distribution of shear stress.

The stress-concentration index (or 'notch sensitivity') of materials

Where notches are concerned, however, the material involved is important as well as the type of stress. Some materials are fairly insensitive to stress concentrations resulting from notches —this applies for example to leather, rawhide and certain kinds of hardwood, which have a texture such that even quite sharp notches have no effect. Others, such as glass, chinaware and high-grade or hardened steels, are particularly vulnerable. This varying reaction to notching has led to the formulation (by Prof. Thum of Darmstadt) of a concept of materials theory, known as the *stress-concentration index* of the material. We know that soft steels are very little affected by notching; stress peaks cannot arise in these steels, for as soon as the stress in the edge areas reaches the yield point, plastic deformation occurs (Fig. 176). Instead of rising to the high value of curve *a* in diagram I, the stress stays, for the most part, at the yield point *c*, as a result of which adjacent areas are more highly stressed, as seen in curve *b*. Since the edge zone bordering the notch (shaded) remains distorted when the stress has been removed, residual stresses appear (diagram II), combined with a very slight increase in overall length. As a general rule, no further disadvantage accrues from either of these, and this yield mechanism thus has a beneficial effect on the strength of machine parts and it explains why many of them do not fracture prematurely as a result of notching.

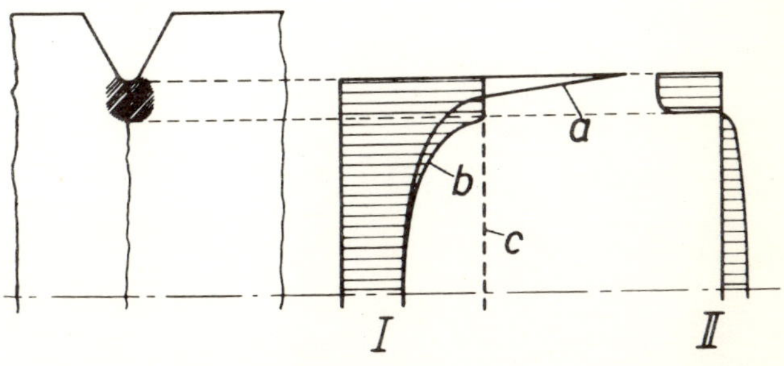

Fig. 176.—In the area directly affected by the notch, yielding will occur with a sufficiently high load: the stress here does not rise above the yield limit, and plastic yielding takes place, as a result of which the load is spread more evenly over the cross-section.

As the grade of the steel rises, so does the limit of elasticity, and the ability to yield drops off; consequently, peak stresses are increasingly less dissipated through plastic deformation, and the result is the high vulnerability to notching and scoring that is so characteristic of high-grade materials. This point must never be forgotten when making use of such materials; although today we have come to expect wonders from alloyed and tempered steels, this is only possible if every kind of notch effect is avoided. The carefree way in which designers turn to high-grade steels if there is failure of mild steel may quite easily lead to a fracture occurring even sooner. One of the best examples of this fracturing is seen in scored glass, which has only a fraction of the resistance in bending of glass with an intact surface.

The interesting thing to note is that the depth of the scratch made in the glass during the cutting process is quite small—so slight, in fact, that we can say that the loss in cross-sectional area has not played a decisive part in the fracture process. The large number of fractures that occur as a result of apparently negligible incisions in the surface of a component are proof that the size of the notch is of only secondary importance in producing a stress peak. Fig. 177 shows the broken end of a shaft from an electric motor, revealing a minute shoulder at the point of fracture which is visible in the photograph as a line running round the circumference. The fact that this shoulder was so small clearly did not prevent it from leading to a local increase in stress which reached the fracture limit. A second example is the big-end bearing cap from a DC-4 engine, seen in Fig. 178, which shows a fissure leading away from a small lubrication passage (in the right foreground) and splitting the whole of the inner surface of the bearing. A crack like this can have very serious consequences, since it means the failure of a substantial part of the bearing area, which is thus likely to break up completely at any time. If ever there was a case where we can speak of a minor cause having major consequences this is surely it for, seen as a part of the whole cross-section, this tiny hole is almost negligible; what is not negligible, however, is the increase in stress which it produces at one point in the overall stress pattern.

Fig. 178.—Bearing cap from an aircraft engine, showing a dangerous fine crack running from a lubrication hole.

Fig. 177.—Engine shaft fractured by prolonged alternating loading, with fracture starting from an insignificant surface discontinuity.

The most convincing proof of the fact that it is the shape rather than the size of a notch or indentation that is important is provided by the fracture of a driving-rod in a piece of sawmill machinery, shown in Fig. 179. This rod suddenly snapped in two during use, producing first of all the fracture surface seen on the left. The appearance indicates a fatigue fracture, starting from the small notch in the top left-hand corner. As soon as this fracture occurred, the rod was no longer located at one end, and began to flail about until a second break occurred at a spot which had also been weakened by a forging flaw (fracture surface seen on the right of Fig. 179). This latter flaw penetrated deep into the cross-section of the rod, but the shape of its outline was relatively harmless, so that the cross-section had been able to cope with normal loading despite the loss of area. Although the area of this second indentation was about 100 times greater than the first it was, as events showed, the less dangerous.

It is, fundamentally, quite easy to understand how the size of a notch is less important than its shape, for the way in which

Fig. 179.—Fracture surfaces on a connecting rod from sawmill machinery. Left, fatigue fracture starting from a small indentation at the top left corner; right, impact fracture occurring only as a result of the previous one, although the cross-sectional area had been more seriously reduced by a casting flaw.

Fig. 181.—Planet gear from an aero-engine drive system, with carefully polished surface, especially at the foot of the teeth.

the lines of force deviate around an indentation in the surface stays geometrically the same whether the notch is enlarged or reduced. What alters with increasing size of a notch is the depth to which it acts—the stress peak becomes wider and extends further into the workpiece, while its intensity remains the same (Fig. 180).

This state of affairs is only altered basically when the size of the notch is comparable to that of the structural elements of the material (which, in the case of metals, means the size of the crystalline grain). This can easily be judged from the grain structure seen at the fracture surface. It provides a valuable criterion for assessing the notch sensitivity—or stress-concentration index—of the material concerned. Coarse-grained materials, such as mild steel and especially cast iron with a high graphite content, should hardly react at all to minor notches and dents, although this does not mean that the internal compensating mechanism is the same in both cases. On the other hand, high-grade, and especially hardened, steels have a very fine fracture surface showing a small grain-size, so that a substantial reaction to notching must be expected. If the fracture surface is flinty or even glassy, then this property is even more pronounced, as illustrated by the example of glass cutting. In fact, fractures regularly occur with hard and brittle materials which have started from apparently trivial surface damage. The scoring which occurs during even fine machining and grinding is quite enough to produce these fractures; they can, however, be almost entirely avoided by careful lapping, honing, burnishing and polishing to produce a smooth surface finish. In this connection we might point to the highly-polished surface of the big-end cap seen in Fig. 178, which has obviously been very carefully finished with this very point in mind; this is, of course, rather inconsistent in view of the dangerous lubricating passage. The care that is expended on the surface finishing of components made from high-grade materials can be seen from the gearwheel shown in Fig. 181 (again from an aero-engine); one is struck by the clean smooth curves in the gaps between the teeth, which as we know are subjected to very high loads and are the points at which most fractures in gearwheels occur.

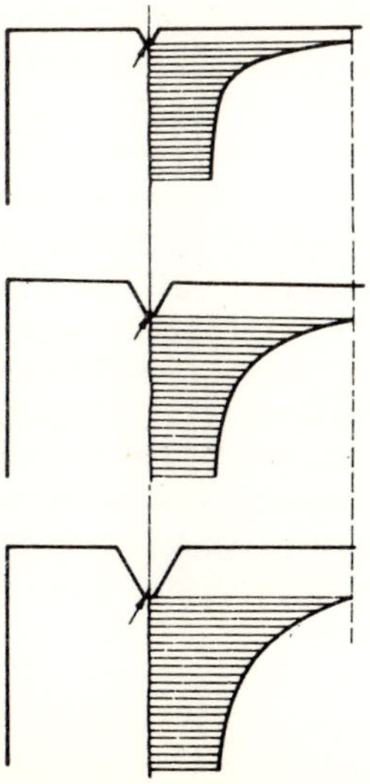

Fig. 180.—Stress pattern for various notch depths of uniform shape; the stress peaks are of equal height but varying width.

The chart in Fig. 182 summarizes the results of experimental work which are surely some of the most impressive findings available on this subject; it shows the decrease in strength of various qualities of material when exposed to surface damage from a number of causes. This chart is of such importance that one is tempted to recommend it as a pattern for the wall-paper to be used in the offices of all leading designers! We need only point here to one feature of this chart—the fact that severe corrosion can reduce the breaking strength of a steel—from $1{\cdot}5 \times 10^9$ N/m² (220,000 lbf/in²) to 13 per cent of its original value—i.e., 2×10^8 N/m² (29,000 lbf/in²), which is barely the capacity of a cheap structural steel under an alternating load. Though it may not always be as bad as this, the loss of strength that results in all materials from only slight surface damage is nevertheless considerable, and it is essential that this be borne in mind when arriving at the cross-sectional area to be used in a design. *If these facts had been available earlier, untold instances of damage to machinery—some of them with disastrous consequences—could have been avoided.* Even today they are not given sufficient attention by leading designers.

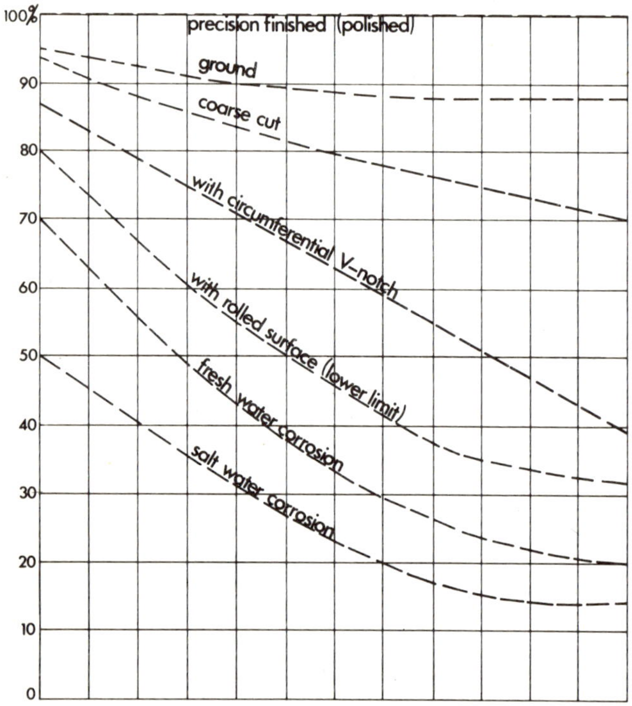

Fig. 182.—Effect of surface finish on the reduction of the resistance to prolonged alternating loads of steels of different qualities (3×10^8 to 15×10^8 N/m²). Most affected are the high-grade steels, where even slight surface flaws lead to a 10 per cent decrease in strength.

Designing to match the stress flow

If there is anywhere that the voice of experience should be listened to and taken seriously, it is here—for this heading covers most of the cases of damage to machinery that occur today. Many designers have also failed to realize that this is a field where real economies in material can be made; they could manage with literally a fraction of the high-grade steels they use if when using these materials they would design so as to match the pattern of the lines of force.

If we look at machine parts from the viewpoint of functional suitability alone, it is easy to see why their surface is covered with so many notches and indentations. Fig. 183 shows the shaft of a heavy-duty lathe. Its purpose, and the way in which it is related to adjacent parts, calls for several variations in diameter, most of which are separated from one another by a recess. Every one of these recesses represents a notch in the surface of the shaft. It is not suggested for a moment that because of these the shaft is likely to snap apart; it needs to be stiff and it has been made so massive that there is no question of any excessive local load occurring. From this point of view the shaft could not be bettered. If it were to be highly stressed, however, its shape would need to be altered in many ways. The example is typical in that time and again it is in the same circumstances that the rules of strength flow are violated. We shall review these rules here, and discuss ways of avoiding such circumstances, or at least of mitigating their adverse effects.

To start with, there is the chamfer, or shoulder, on a shaft. This serves as a bearing surface for gears, levers or bearings and, as it often has to take fairly considerable axial loads, it cannot be made as narrow as we would like. Fig. 184 shows the best ways of shaping such a chamfer. First, the dangerous, sharp corner (*a*) is replaced by a curve (*b*). In addition to this, it is useful to give the shoulder a sloping surface (*c*); particularly advantageous, however, is a sloping surface combined with a curve (*d*), for we can see that with the same width of bearing surface the radius of curvature can be very much larger than with the right-angled shoulder. Measures like these are likely to reduce the notch effect index by several units.

If such measures are not sufficient, then we make use of *stress-relieving notches*, by means of which we can influence the pattern of the lines of force from without. Looking at the shoulder machined into the shaft shown in Fig. 184*a*, we can see that what makes it particularly dangerous is the fact that it cuts abruptly into the lines of force which, coming from left to right, run through the whole cross-section of the shaft and are suddenly pinched together. This would not be the case if the change in cross-section were to occur gradually (as indicated by the dotted line in Fig. 185*a*). Then, however, we

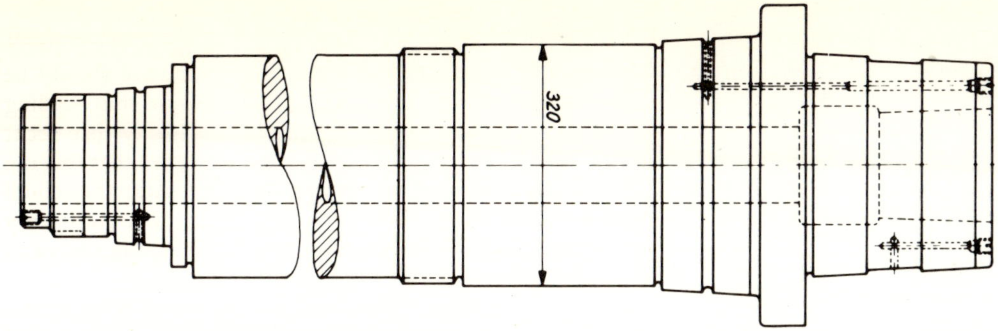

Fig. 183.—Large lathe shaft as an example of how functional requirements can lead to the surface being broken up into a large number of separate shapes, most of which have a notch effect.

would lose the shoulder as a bearing surface, so we have to arrive at the compromise seen in Fig. 185*b*, where a notch has been made alongside the shoulder which will gently squeeze the lines of force together and guide them through the sudden constriction. Functionally speaking, the only purpose of this second notch is to relieve the stress on the first, and we therefore call it a 'relief notch'. The form shown in Fig. 185*b* is, however, an ideal and in practice we can only achieve an approximation to this (as shown in Fig. 185*c*). Since the point of these relief notches is not properly understood by many people, however, we often come across examples like that in Fig. 185*d*, where the relieving effect is of course insufficient, or in Fig. 185*e* where there is only a very slight improvement.

Screw threads of all types can be looked upon as a long series of notches, with each individual turn of the thread losing much of its dangerous effect through being screened and relieved of stress by its neighbours, along the lines explained above; but this does not, of course, apply to the turn which is nearest to the approaching lines of force. Consequently this part of the thread needs the protection of a proper relief notch if it is not to lower the strength of the machine part concerned. Fig. 186 shows the most effective way of providing such a relief notch.

In cases where it is impossible to provide a relief notch of this kind, the thread profile should be modified so that the valleys of the thread compress the lines of force as little as possible, i.e. the sides of the thread should have a moderate angle, and the bottom of the valley should be well rounded (Fig. 187).

However, a screw thread does not only present a problem in tensile stress—where it is inside a nut, the thread is subjected to a bending load as well. Consequently there is always an extra tensile stress deep down in the thread groove, the effect of which is to locate the most vulnerable section not at the point of transition between the thread and the cylindrical body of the screw or bolt, but inside the nut; and this will be concentrated on the first turn inside the nut, since this is where most of the load is carried. Any carelessness in shaping the bottom of the thread groove will take its toll, and this forces us to adopt very painstaking finishing methods such as milling, grinding or even burnishing. In the last case, the bottom of the thread groove is rolled, cold-formed and hardened with a hardened and polished steel roller after being cut with a tap or on the lathe. The pressure used is so great that plastic deformation takes place and a residual compressive stress remains when the roller is removed. The inside of the cross-section is thus placed under a tensile stress (Fig. 188, diagram I). If this

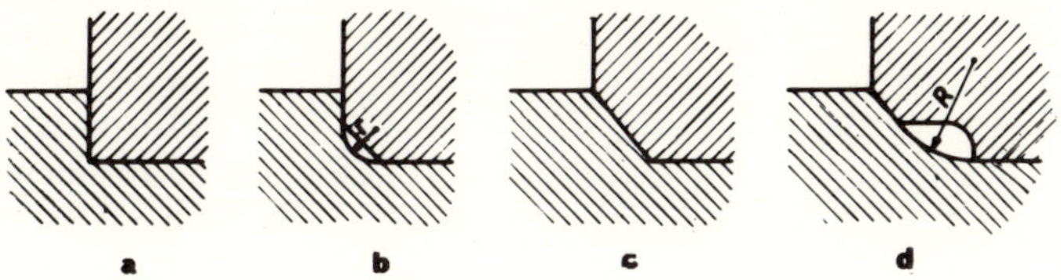

Fig. 184.—Shaft chamfers can be given a wide variety of shapes without affecting their functional suitability. Sloping surfaces, combined with a radius (example *d*), are particularly suitable.

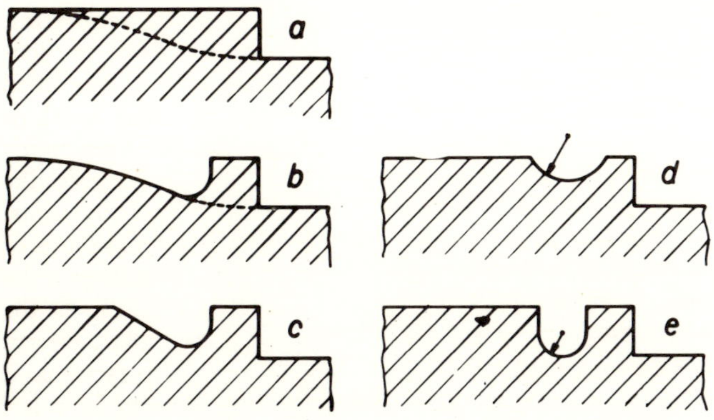

Fig. 185.—Guiding the lines of force by means of notches which are not functionally essential is a useful method of reducing the detrimental effects of notches that cannot be avoided. These are termed *relief notches*.

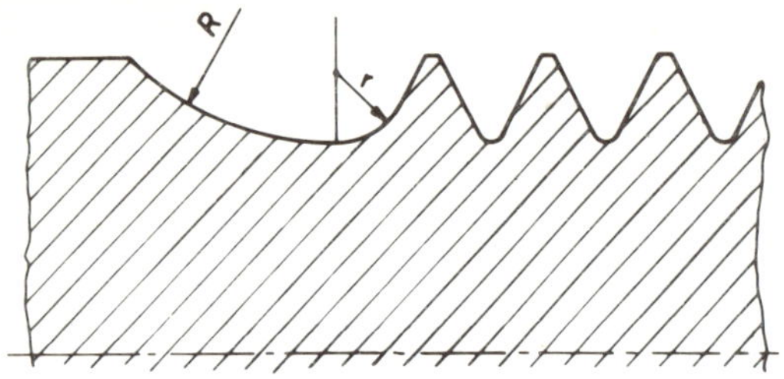

Fig. 186.—Relief notch where a screw thread meets the cylindrical body of the bolt.

stress pattern is now superimposed on the normal tensile pattern (diagram II), the result is a pattern as shown in diagram III, where the dangerous stress peak at the base of the thread groove has been very greatly reduced. The excellent practical behaviour of this type of thread provides ample proof of the correctness of the procedure.

Fig. 187.—Ideal thread profile, with 45° angle and large radius at bottom of groove, giving a good strength flow pattern.

Drilled holes are a common feature of machined parts—they are required by the function of these parts, and are intended to accommodate screws, pins and the like and sometimes to serve as passages for lubricant and coolant fluids. Fig. 189 shows a hollow eccentric shaft with, at each bearing, one or two radial holes intended to act as an outlet for oil passing down the centre of the shaft. Although only a small proportion of the cross-section is lost by the drilling of these holes, they bring about a substantial weakening of the shaft through their notch effect. The bearing cap in Fig. 178, too, shows how easily oil holes of this kind can lead to a fracture. The piston rod in Fig. 190, which has several holes in the sides to allow the coolant to reach the piston, was evidently dismantled only a very short time before it was going to fracture; the innumerable small cracks inside the crosswise holes are a clear proof that this was imminent.

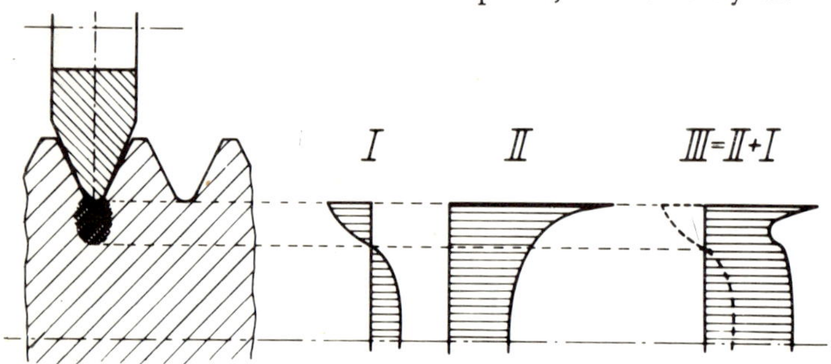

Fig. 188.—Cold-forming of the bottom of the thread groove following the cutting or milling operation is useful; it produces a residual stress at the most vulnerable area which reduces stress peaks which would otherwise be dangerous.

In all these cases, however, holes like this are a convenient answer to a functional problem, and it would be difficult to do without them entirely. Our previous example of the thread which was cold-formed at the bottom of the groove shows us one way out of the difficulty. Smooth, hardened steel arbors are driven down into holes of this kind so that there is an area of plastic deformation on the internal surface, with a residual stress pattern which reduces the concentrations of the main stress to a tolerable level (Fig. 191).

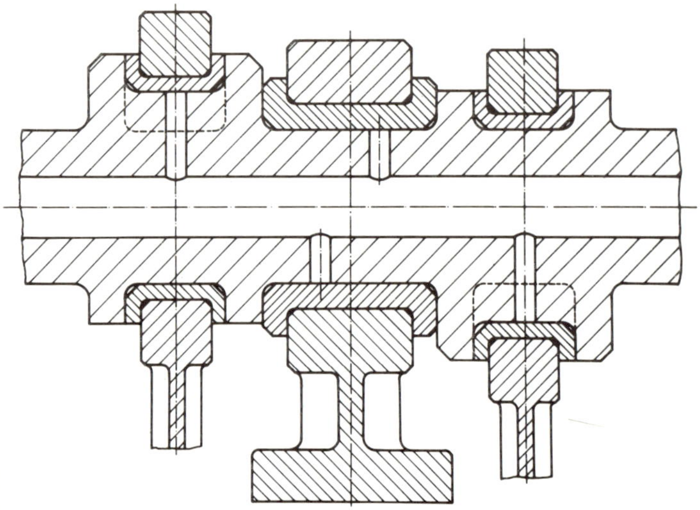

Fig. 189.—Many machine parts have small holes drilled in them for a wide variety of purposes. Here an eccentric shaft with a central oil passage has small radial oilways drilled in it for lubricating the bearings.

A further example is provided by *keyways*, which are typical of a certain type of recurrent design fault. Using Prandtl's 'soap-bubble' analogue* it can be shown that even with torsional loading every indentation in a profile will lead to a stress peak the level of which will rise with the abruptness of the indentation; it is thus obvious that a keyway should never be made with sharply-angled sides. Parts made from high-grade materials and subjected to alternating loads are particularly vulnerable. Fig. 192 illustrates a typical fatigue fracture in a shaft, starting from the corner of a keyway. As will be seen, this corner has been generously radiused but even this was unable to prevent an overstressing of the material at this point. Abrupt edges are even more likely to lead to fractures.

* This states that the distribution of shear stresses over a section undergoing a twisting force is proportional to the inclination of a membrane stretched across this section and subjected to a pressure from one side.

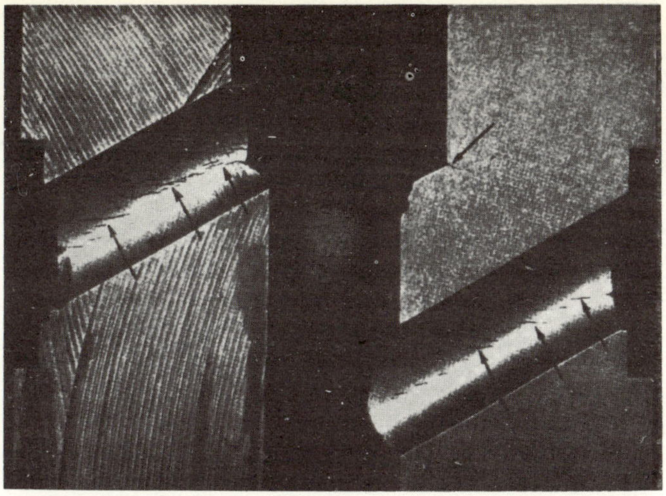

Fig. 190.—Piston rod weakened by small transverse oilways, showing fine cracks well on the way to causing a fracture.

In Fig. 193 we see the broken end of a splined shaft from a woodworking machine. The break probably started at the bottom left, from the corner of the keyway at that point, and then ran spirally round the shaft until it reached the corner groove of a flange coupling, whereupon the whole end finally became separated. A second crack ran from the uppermost spline from right to left, so that the missing triangular part dropped out. It is not surprising that breakages and breakdowns occur again and again when the profiles of splines and keyways are made with this kind of sharply-angled corner. Even the radii of curvature quoted in standard specifications for keyways (e.g. B.S.4235: Part 1) are barely adequate. Fig. 194*a* shows a drawing, to scale, of a keyway measuring 18×8 mm for which the Swiss VSM specification 15116 gives a corner radius of 0·25 mm. In comparison with this the radius of the keyway in Fig. 192 seems generous. There is no doubt at all that this point right at the bottom of the groove is by far the most highly-stressed area of its whole cross-section, so that the height of the load-bearing side of the groove could quite easily be reduced to make way for a greater radius. Reducing it by one-third would raise the pressure of the key on the side of the keyway by 50 per cent, but would allow the corner radius to be increased from 0·25 mm to 3 mm; this is shown to scale in Fig. 194*b*. The result would be to cut down the notch effect not just by a few per cent, but to a small fraction of the original value. This is still not to say that the detrimental effect of the groove will have been overcome; as a rough estimate, the stress-concentration index of a semi-circular groove on the periphery of a shaft would be in the region of 2 to 3. So rounding the corners at the bottom of a keyway can only be to the good, and it is difficult to understand why designers are so loath to make use of it when planning this highly-important detail.

Special attention ought to be given to providing a generous radius in the case of spline shafts, which as a rule are even more highly stressed. When a fracture occurs in one of these shafts, it nearly always runs from the base of one of the splines, even when the radii here are quite large. Fig. 195 shows the fracture surface of a ten-spline shaft. In spite of the raggedness of the surface, we can see quite clearly how the cracks started from the corners of the grooves and ran towards the centre. It is obvious that these cracks are connected with the fact that the part shown came away; but from the other side of the

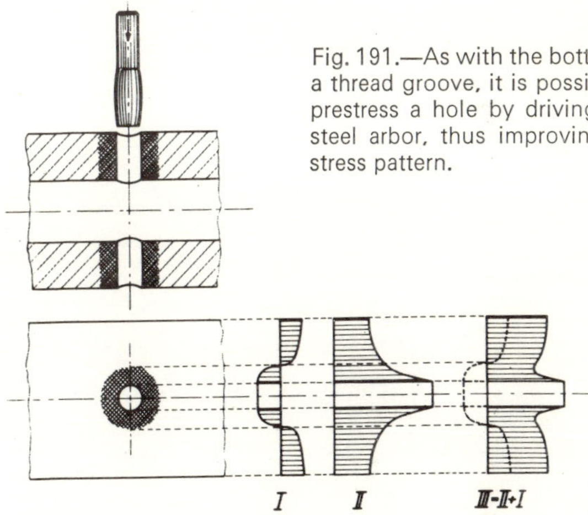

Fig. 191.—As with the bottom of a thread groove, it is possible to prestress a hole by driving in a steel arbor, thus improving the stress pattern.

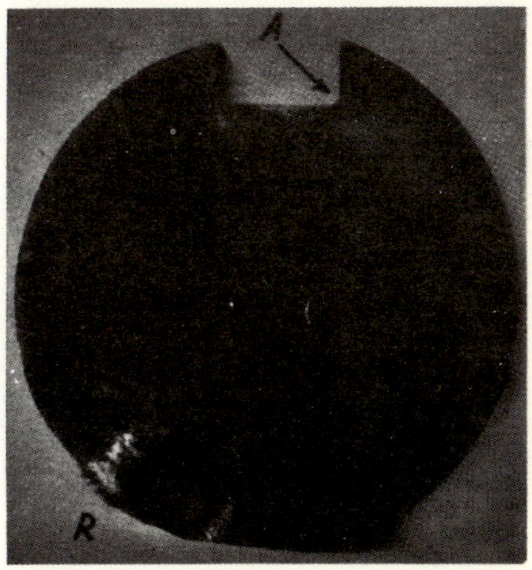

Fig. 192.—Stress peaks around sharp angles at the bottom of keyways are a frequent cause of fractures in shafts. Starting at A, the fracture finishes at R.

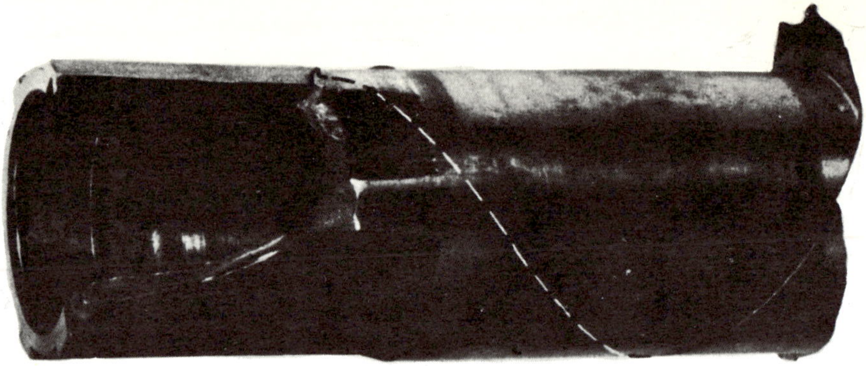

Fig. 193.—Shaft with a sharp-edged spline profile, where high stresses at the base of the spline were the obvious cause of fracturing. The fracture runs spirally (dotted line) to the recess on the right.

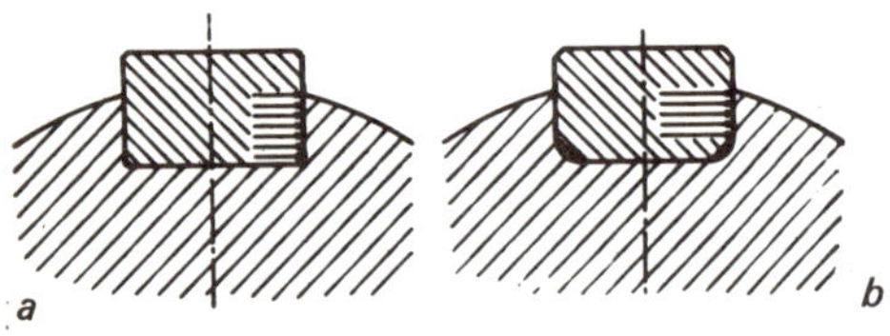

Fig. 194 *a* and *b*.—Keyway profiles to VSM standard: very small radii at bottom of groove weaken the shaft. Profile (*b*) is better, giving improved strength despite reduced bearing area through much larger radius at bottom of groove.

broken part (Fig. 196) it can be seen that they occurred not only in the immediate neighbourhood of the break, but also penetrated deep into the material of the shaft in the axial direction. We can also see how far the cracks run in the radial direction, some of them even reaching the centre of the shaft. All these signs show that the separation of the section shown was only part of a more extensive fracture process which was literally breaking the shaft into small pieces. The immediate cause of this must, again, be identified with the inadequate radii given to the base of the spline grooves. If, however, radii like these and those seen in Fig. 192 have proved in-

Fig. 195.—Fracture surface from the rear axle of an earth-moving machine, with the fracture starting from the sharp corners at the bottom of the splines.

Fig. 196.—Other side of part of axle shaft seen in Fig. 195, showing that cracks had started from one corner of each of the splines, penetrating deep into the metal in both the axial and radial directions.

adequate, then all shafts where the grooves have been given little radius or none at all must be treated with the greatest distrust. To sum up, we may say that keyways and splines can never be given too generous a radius at the corners of grooves, and that it is no exaggeration to shape a spline shaft as shown in Fig. 197. The profile with smaller radii shown as a dotted line does, admittedly, have a 20 per cent greater moment of resistance; but this advantage is many times outweighed by the substantial notch effect resulting from the sharp transition from wall to base. The shaft with the larger radii is thus very much stronger, besides being about 12 per cent lighter.

Fig. 197.—Large radii at the bottom of splines can only be to the good. The dotted lines show spline profiles to standard specifications; in spite of having a higher moment of resistance, the latter are not as strong.

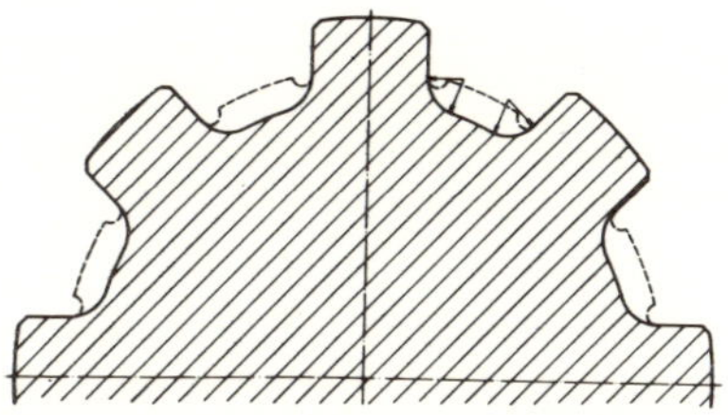

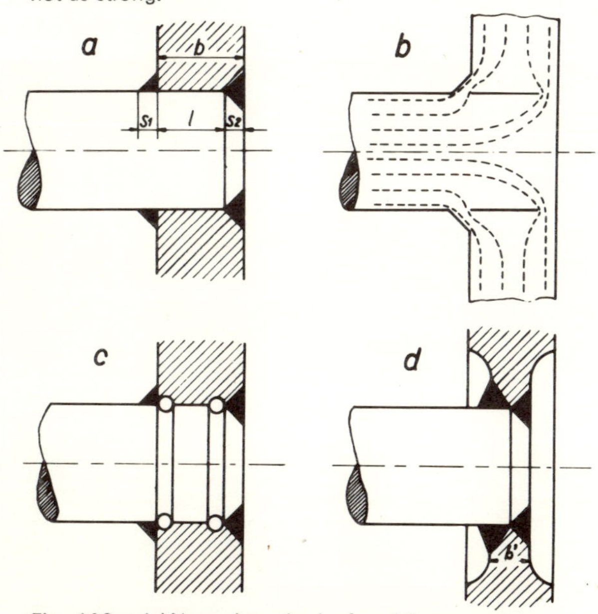

Fig. 198.—(*a*) Normal method of welding a rod to a round plate.
(*b*) Strength flow pattern with interface notch inside workpiece.
(*c*) Attempt at reducing the notch effect by providing grooves to remove sharp corners of notch.
(*d*) Interface notch avoided by carrying weld metal right through the reduced thickness *b*′.

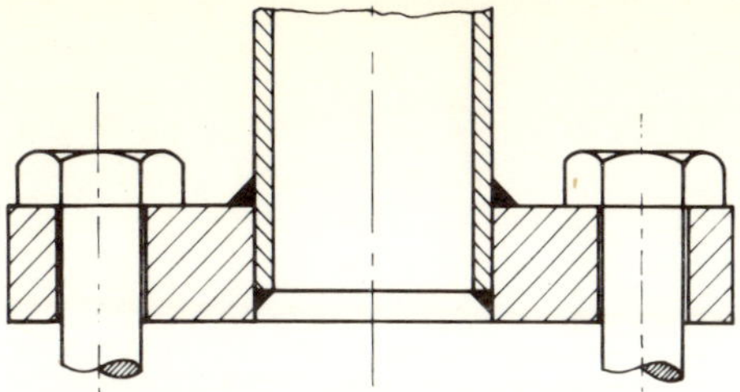

Fig. 199.—Common method of joining tube and flange by two weld beads, leaving an interface notch between them. Starting-point for cracks, especially if subject to vibration.

Notches may also be concealed within a workpiece in the form of cavities, inclusions and forging flaws; but since the occurrence of such defects is beyond the direct control of the designer, they cannot be termed design faults. There is, however, one kind of internal notch which must be looked upon as a design fault; it is the direct result of certain types of welded joint where, in making the weld seam, a particularly dangerous situation known as an *interface notch* can arise. Fig. 198*a* shows what is meant; here the shaft and circular plate are in contact over a certain length l, to which is then added the weld beads s_1 and s_2. The way we have indicated our welds hides the fact that an interface notch exists over this length l. This fact is, however, 'felt' by the lines of force as shown in Fig. 198*b*, where the danger of the situation becomes very clear. It must be admitted that joints of this kind are very common indeed in engineering practice—to quote only one example, flanges are frequently joined to the ends of tubes in this way (Fig. 199), and joints of the type seen in Fig. 200 are also very common. Naturally, fractures are not long in appearing. The cracks spread from the middle into the weld

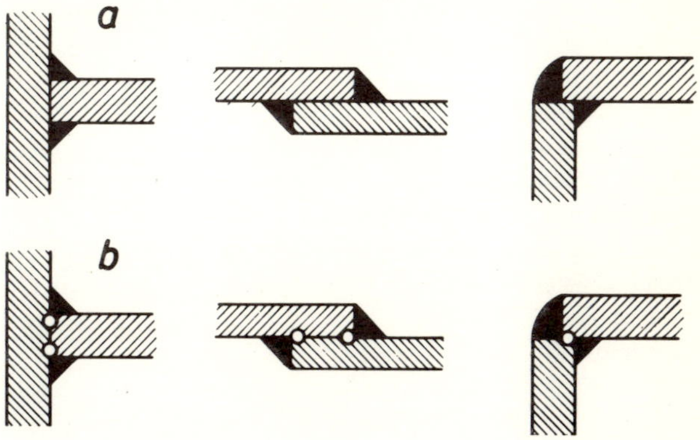

Fig. 200.—(*a*) Types of joint often used in welding practice, all of which involve interface notches.
(*b*) Improved joints through boring-out sharp edges.

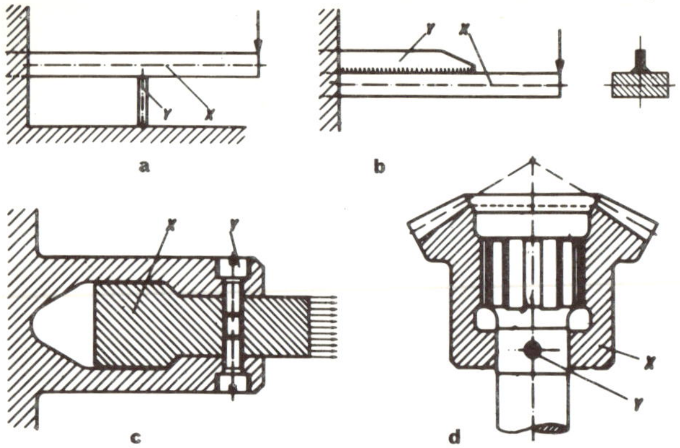

Fig. 201—Examples of systems where distortion is prevented from taking place.
(*a*) Bar under flexion, with small support under compression intended to relieve load.
(*b*) Bar under flexion with added spline, which only appears to lend extra strength.
(*c*) Tie bar attachment, with locking bolts to prevent bending of lugs.
(*d*) Pinion of spline shaft, locked axially by pin.

bead, where they run to just below the surface without their presence becoming visible from the outside.

Here, again, we must look for ways of avoiding or reducing the harmful effects. The first of these might be to drill away the sharp edges of the notch (Figs. 198*c* and 200*b*). It is immediately obvious, however, that this will only be possible in cases where the whole joint does not have any great depth perpendicular to the plane shown in the diagram. With joints like that in Fig. 198, circular grooves could be made before making up the joint, so that the notch effect is somewhat alleviated. But far better than any of these measures is to combat the evil itself by making the weld beads continue right through from one side to the other, extending the welds s_1 and s_2 until the length l disappears altogether. This does, however, assume that the width b is not too great, although this can be dealt with by making a slight change to b' (Fig. 198*d*).

Finally, we must mention a type of design fault which as such would not be too difficult to recognize if it were not very often hidden from the designer's eye by particular circumstances. This is to be found in mechanical systems which are prevented from undergoing a distortion proper to them by some apparently indispensable additional component. A few examples of this are shown in Fig. 201, where the main system is labelled x and the additional part preventing the distortion is marked y. In cases (*a*) and (*b*), the distortion is in the form of a bending, in (*c*) a longitudinal tension, and in (*d*) a twisting action. The elements marked y act in opposition to this distortion and since—as we can see from innumerable practical examples—they are very often not made strong enough to do this effectively, a fracture results. Thus in (*a*) the support will buckle, in (*b*) the welded-on strip will tear apart, and in (*c*) and (*d*) the bolts and locking-pin will shear (resulting in the whole assembly coming apart). The attempt at strengthening a beam by adding a knee-plate seen in Fig. 156*c* is a further example of this type of design fault.

There is no doubt that an accurate calculation of the strength of the whole system would show immediately the subsidiary part y has been made too small. The unfortunate thing is that this kind of problem is always statically indeterminate and is therefore difficult, if not impossible, to calculate. In most cases, indeed, the situation is not even recognized as being a difficult one—the fact that the additional part y has been positioned where substantial stresses must arise as soon as distortion is prevented from occurring is simply overlooked. If the designer only got as far as recognizing this, he would appreciate the danger and probably make the additional part large enough to obviate the risk of fracture.

Fig. 202 shows a very popular method of building up armatures, which represents a pitfall for any designer who is not aware of the dangers just outlined. It frequently happens

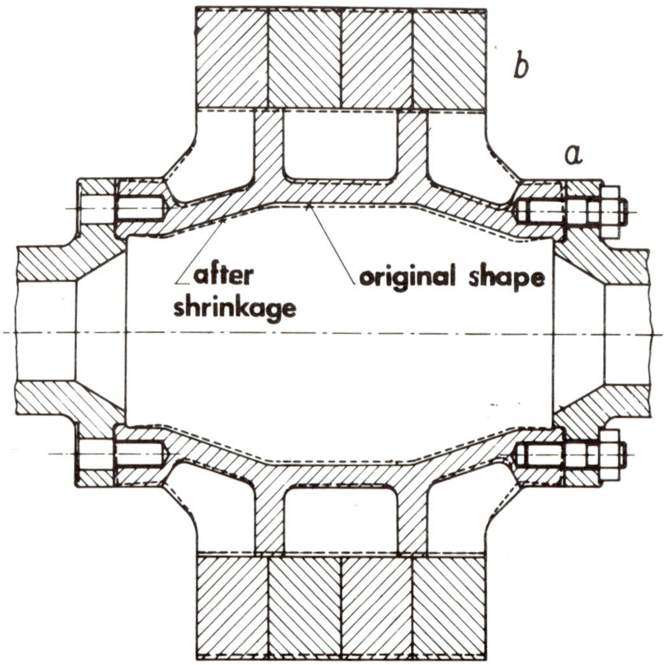

Fig. 202.—Popular modern method of assembling armatures for electric motors. When the rings are shrunk on, the wheel centre flexes inwards and the bolted-on side flange is canted. This exerts unexpected extra loads on the bolts.

Fig. 203.—Knee-plates added to corner joint of a cable-railway bridge structure to provide extra strength (Central-ETH, Zurich). This is not good practice. See text.

that the bolt attachments *a* for the shafts on either side are snapped off when the pole rings *b* are temperature-shrunk in place, because they try to prevent the heat distortion from taking place. The wheel centre contracts as the rings are shrunk on, and the inner flange moves slightly as indicated by the dotted lines. If the bolts are not strong enough to prevent this distortion, or not elastic enough to allow it to occur, then they will snap. Fig. 203 shows an example from girderwork practice, where knee-plates have been welded into the corner joint between vertical and horizontal beams, obviously with the idea of making this important joint stronger. It will be plain, from what we have seen above, that this is something that cannot be done effectively in this way; only if the plates were several times thicker would there be any real gain in strength. The fact that a fracture has not occurred is due only to this joint not having so far had to withstand any really high loading.

Fractures of this kind are particularly common with torsional loading. Substantial distortion is not expected in such cases—a reasonable assumption provided that we are dealing with a closed section which (as we have already seen) has exceptional torsional resistance (cf. Figs. 133 and 136). Open sections are quite a different matter, however, and unless this is kept in mind trouble will occur. We see an example in Fig. 204, where we have a star-shaped open section *a* with encircling rings *b* attached to it at regularly-spaced intervals. When the assembly is subjected to a torsional load the section will twist a good deal because of its low torsional resistance, and the outer edges, which were originally straight lines, will

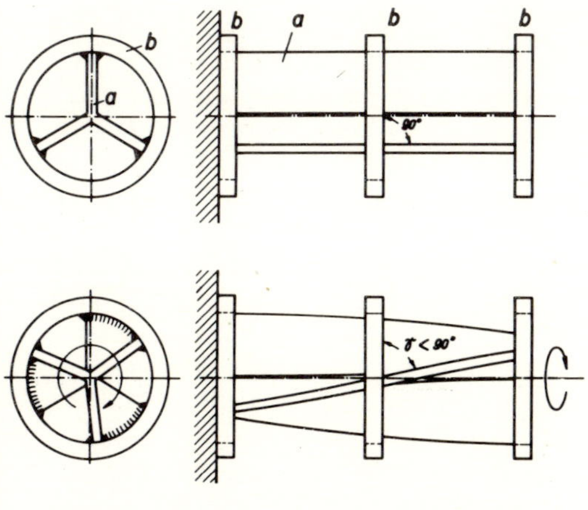

Fig. 204.—Torsion bar with open section and low torsional stiffness, with attached rings. Top: unstressed. Bottom: under torsion with the rings trying to prevent the bar from twisting.

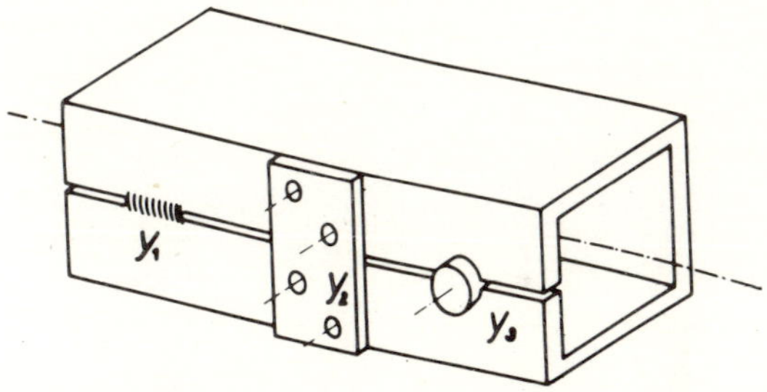

Fig. 205.—Square tube, made into open section by lengthwise slit joined across at individual points y_1, y_2 and y_3.

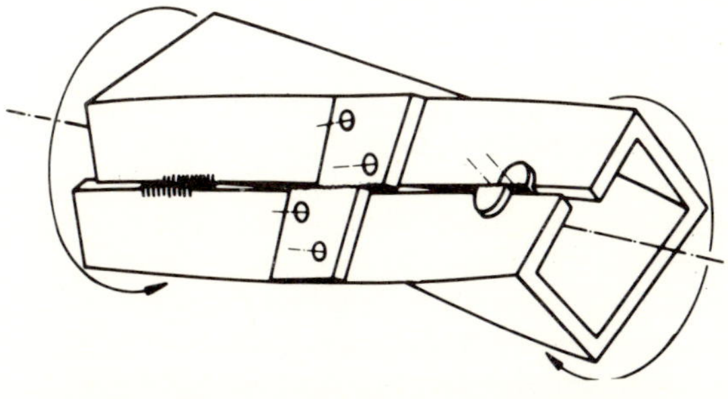

Fig. 206.—When the tube is twisted, the edges of the slit move with respect to each other, and the links *y* are subjected to shear stresses which may cause them to break apart.

become helixes. These helical lines are oblique to the main axis, while the rings will remain at right angles to it and thus try to prevent the twisting; if and where fractures will occur in a system like this depends entirely on the dimensions of the parts concerned.

Similar unpleasant surprises can occur when a section subjected to a twisting load (Fig. 205) has various kinds of join, y_1, y_2, y_3 added across a lengthwise slit. If these are not sturdy enough, they will shear or tear apart as soon as the girder starts to twist. Fig. 206 illustrates the shape the girder will try to take up; it can be seen that the two edges of the slit

Fig. 207.—In nearly all trucks the load-carrying chassis is made up from two U-section beams facing each other and joined by cross-members.

will shift with respect to each other, and that anything trying to prevent this happening will be subjected to a shearing load.

Our temptation might be to look upon considerations of this kind as being of only academic interest—and yet designs are constantly being produced which incorporate precisely this kind of problem. To take only one example out of many, let us look at a design that is a widespread favourite among designers of vehicle chassis of all kinds. The chassis is made up from two U-section open-sided girders, arranged with the open sides facing each other and joined by a number of separate cross-members, as shown in Fig. 207. If a chassis frame like this is twisted about its lengthwise axis (which is

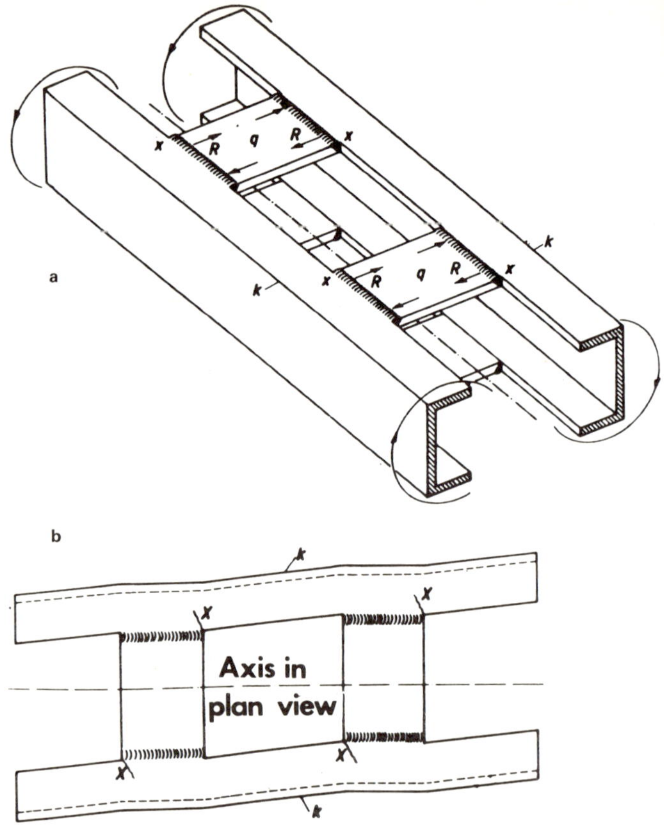

Fig. 208.—(*a*) U-sections with cross-members intended for spacing and stiffening, but which prevent distortion when under torsional loads.
(*b*) Plan view of distorted system seen in *a*. The edges *k* become kinked, and there are high bending stresses at points *x*, at which the sections can easily tear, or the welds may break.

quite possible through the truck being loaded on one side, running over uneven ground and so on) the longitudinal beams twist as shown in Fig. 208*a* and the edges *k* which have so far been straight become helices. This twisting movement is opposed by the cross-members *q* which apply a force R to the longitudinal beams. These forces are concentrated at the corners of the joints *x*, producing stress peaks which very easily break down the welds (if in fact the side-members themselves do not crack at these points, as indicated in Fig. 208*b*). Designers cannot be dissuaded from using these facing U-sections as main chassis members; but they ought at least to arrange the cross-members so that they do not stop the main longitudinal members from twisting; instead of flat plates, they should use vertical webs (as in Fig. 209*a* which can easily

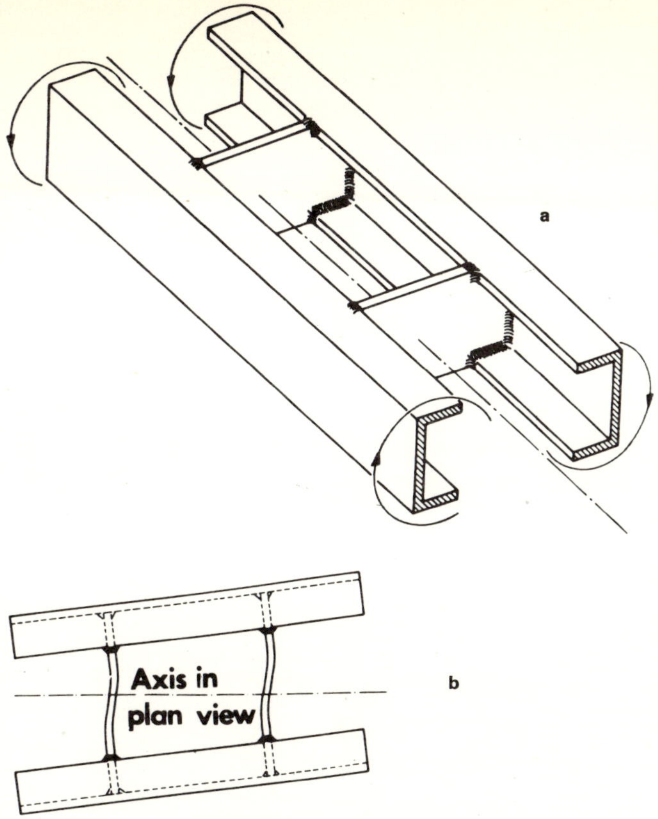

Fig. 209.—(*a*) The same system as in Fig. 208, but with flexible cross-members which can accommodate the twisting of the U-section side members. There is no resistance to the distortion, and the U-sections twist evenly.
(*b*) Plan view of *a*, with the bending of the cross-members allowing continuous distortion of the side members.

Fig. 211.—Merk-Pullax tractor built by the Hans Merk Maschinenfabrik Dietikon-Zurich.

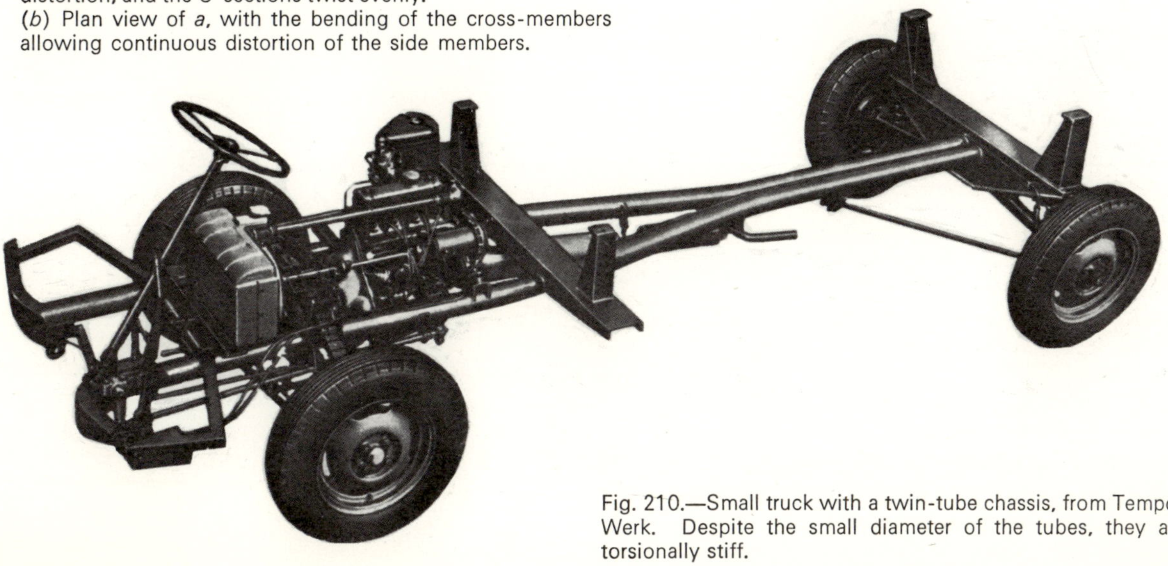

Fig. 210.—Small truck with a twin-tube chassis, from Tempo-Werk. Despite the small diameter of the tubes, they are torsionally stiff.

bend and accommodate the movement of the main members (Fig. 209*b*). Such a design may well give the impression of not being as strong as those we have just seen—but appearances are deceptive, for when we are dealing with the strength of a system it is not distortion but the maximum stresses that arise that count; and with the arrangement seen in Fig. 208*a* the stresses at points *x* are definitely higher than anywhere in the system of Fig. 209*a*, even though in the latter the total amount of distortion may be greater.

However, the best solution of all for design problems of this kind is to replace open sections with closed ones. Because of their great torsional stiffness, these can cope with cross-members of any kind, and the problems we have been looking at do not arise. These closed sections can be produced by welding plates across the opening of a U-section, although round tubing (Figs. 210 and 211) is also to be recommended. What is especially noticeable about the vehicle in Fig. 210 is the slimness of the two chassis tubes—proof enough of the choice of a closed section being the right one. It may seem a 'far-out' idea to use a tube as the load-bearing member for a lorry, yet there is no apparent reason why such a design should not be generally adopted.

Summing up, we could say that the art of good form design is to lead the forces from one point to the next by the most direct route possible, and also to ensure that the lines of force are not constricted at any point; this means that notches must be avoided—anything that interrupts the smooth flow, and in particular the even development of distortion. Equal loads should be carried by equal cross-sections—in other words (to come back to a phrase we have already used) designs should be *homogeneous*. It must be constantly borne in mind that every stress peak that can be reduced makes it possible to raise the overall level of stresses; and this means either that the part in question can be subjected to a higher load, or that it can be made thinner and therefore lighter.

The shaping of corners and surfaces

It should be obvious from what we have just been discussing that corners and edges occupy an important place in design theory, and this is also evident from an examination of any machine part (such as the cylinder cover of Fig. 212) incorporating nearly all the kinds of edge and corner that occur in engineering practice. We have already seen that abrupt indentations in the profile should, as a basic rule, be avoided (i.e. all recessed corners must be given a sufficiently large radius: the larger the radius, the greater the strength).

It is not always easy to follow this rule with welded constructions. Fig. 213 illustrates the various shapes a weld bead along a recessed corner can take. Shape *a* is obviously unsuitable for producing a good strength flow, and *b* is very little better: *c*, on the other hand, is the ideal shape, and whenever strength is an important consideration this is the shape that should be

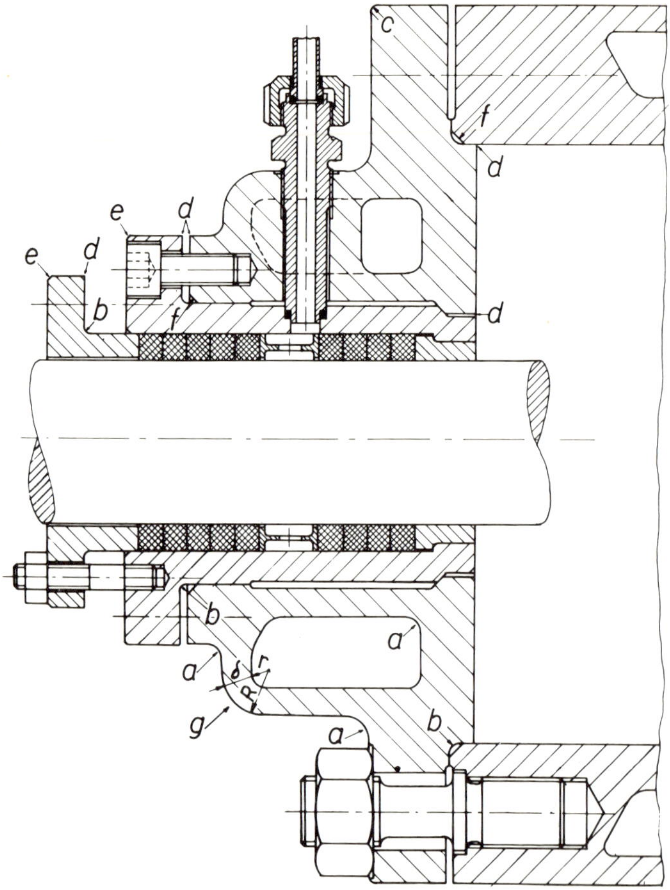

Fig. 212.—Cylinder cover with cooling jacket and gland, incorporating nearly all the types of edge and corner common in engineering practice:
a recessed, rough casting, with large radius
b recessed, machined, with small radius
c projecting, rough casting, with small radius
d machined, covered, left sharp
e machined, free-standing, cut back
f machined, facing a radius, cut back.

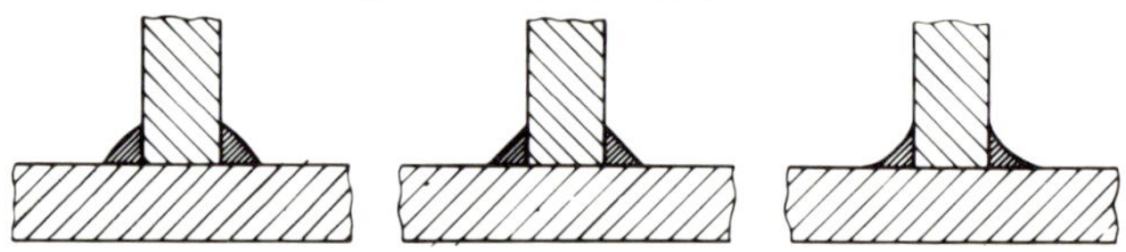

a **b** **c**

Fig. 213.—Weld beads along a recessed edge: *a* convex; *b* level; *c* concave.

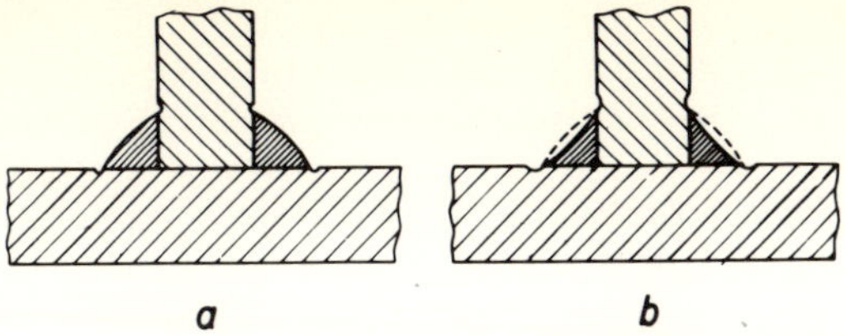

Fig. 214.—
a Weld bead with undercuts;
b concave shape with undercuts ground away.

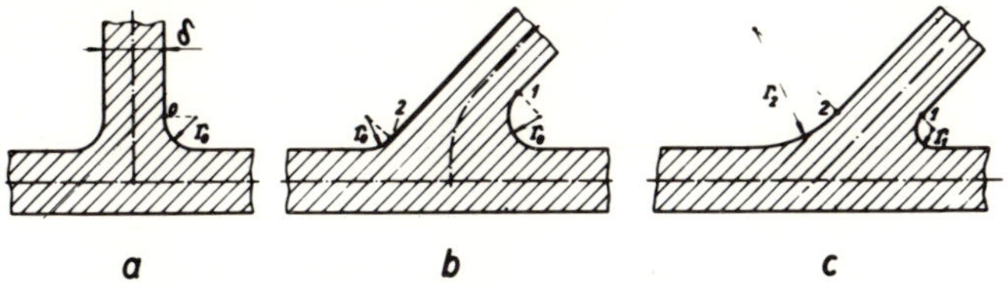

Fig. 216.—Transitions on casting:
a at right angles
b oblique, same radius both sides
c oblique, differing radii.

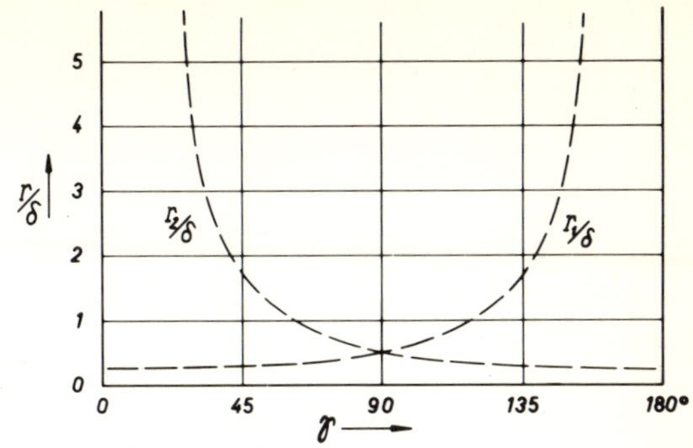

Fig. 217.—Relationship of radius of curvature to angle of attachment.

specified. The fine texture of the surface is also an important factor—if the welding technique is not perfect, the surface will always tend to be flaky and marked by tiny fissures. We have seen that these can have just as detrimental an effect on the strength flow as large notches, and in some circumstances

Fig. 215.—Welded pulley with machine-turned weld bead.

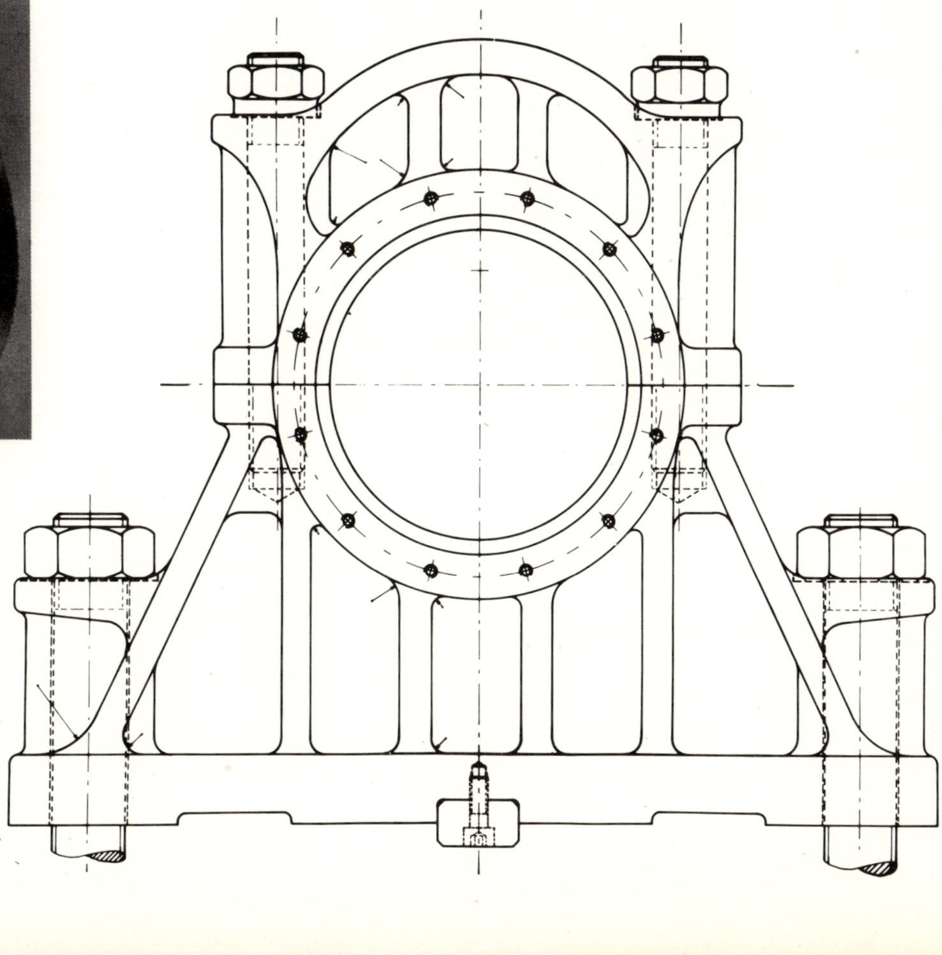

Fig. 218.—Heavy bearing block, an example of graded radii at transitions.

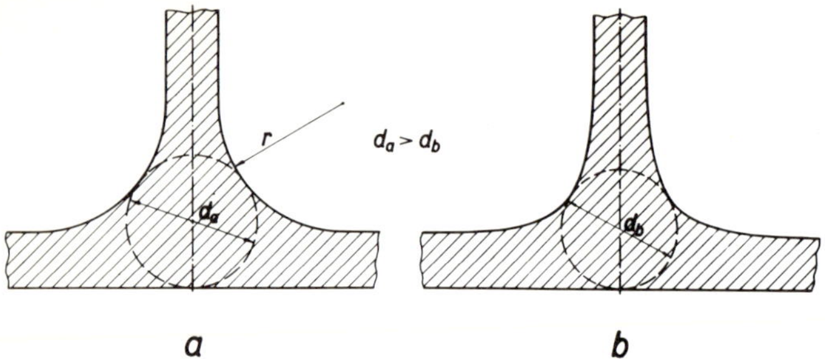

Fig. 219.—Large radii lead to massing of material:
(*a*) with arc to circle
(*b*) with hyperbolic curve.
Diameters d_a and d_b are a measure of the massing of material.

a concave weld bead may be hardly better than a convex one. Added to this there is the fact that with nearly all welds there is along the edge of the bead, where the molten part of the basic material begins, a small but still quite dangerous penetration notch, known as the *undercut* (Fig. 214*a*).

All these features show the weld bead in a very unfavourable light as a part of the shaping of the transition between one surface and the next, and lead us to regard welds in general as no more than coarse surface forms which will need surface finishing if they are to conform to the rules of good strength flow. The first type of finish treatment that springs to mind will be machining away part of the weld metal, or grinding with an emery wheel (Figs. 214*b* and 215). Very much more effective, however, would be hot rolling; this not only gives a smooth surface finish, but also strengthens the whole structure of the weld metal and closes up any surface fissures under the pressure. This may seem like adding complications to our welding technique; yet machining and grinding, too, involve extra trouble and cost and do not give anything like as good a result as roller burnishing. If we think of the expenditure involved in mechanical and metallurgical aids—the UP process (under-power welding), to name only one—then suggesting this method of improving the weld bead does not seem quite so out of the way.

When dealing with iron castings, adapting the shape of transition areas to the rules of good strength flow is a great deal easier, and the designer is able to specify whatever radius he thinks best for the recessed corners (*a* in Fig. 212). Large radii are always an advantage, 'large', that is, in comparison with the wall thickness δ at the point in question. Normally, we adopt the formula $r_0/\delta = 0{\cdot}5$, which corresponds to Fig. 216*a*. This, however, only covers a case where the rib meets

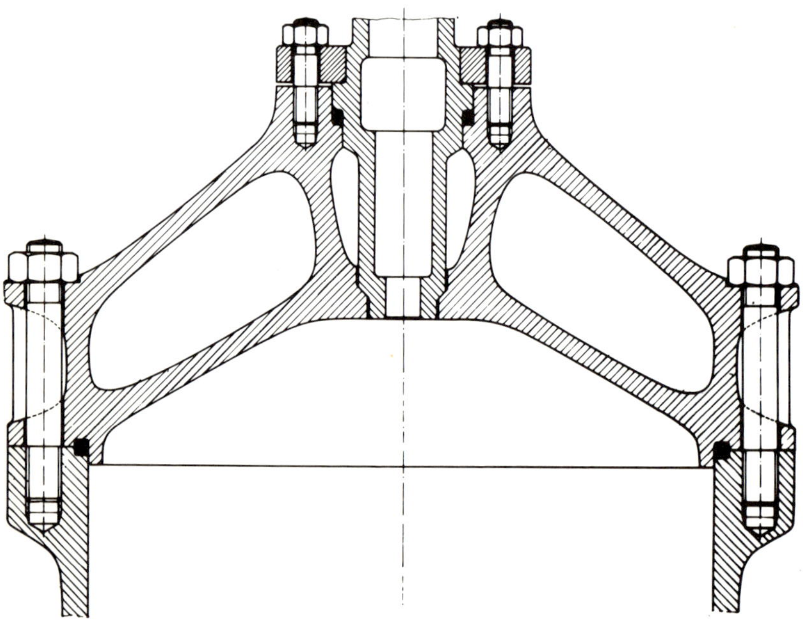

Fig. 220.—Pressure vessel lock carrying high loads, with transitions matched to the stress flow.

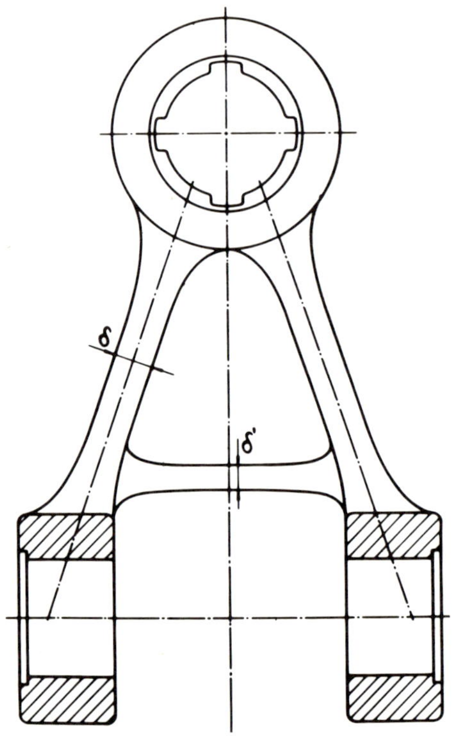

Fig. 221.—Guide fork, with transitions similar to those of Fig. 220.

the base at right angles. If the joint is at an angle γ, we find that using the same radius of curvature on both sides will give a poor transition which makes the axis of the rib appear to be bent just before the junction (Fig. 216*b*), which is not intended. The impression clearly comes from the fact that points 1 and 2, at which the curve joins the straight part, are not the same distance up the rib. A far better appearance is achieved if points 0, 1 and 2 are all at the same height, as in Fig. 216*c*. It is easy to see that this rule leads to calculable relationships between r_1/δ and r_2/δ and the angle γ at which the rib meets the base. These are:

$$\frac{r_1}{\delta}=\frac{0{\cdot}5}{1+\cos\gamma} \quad \text{and} \quad \frac{r_2}{\delta}=\frac{0{\cdot}5}{1+\cos(180^\circ-\gamma)}$$

Fig. 217 defines the shapes of these two curves, which are of course symmetrical about $\gamma=90^\circ$. Naturally no designer will follow this rule slavishly—it is only a general guide, aimed primarily at producing a result that 'looks right' (Fig. 218).

When a design involves very high stresses, these radii of curvature will not be large enough. Even with extreme radii, there will still be increased stresses at these points; so in fact the radii can never be too large.

And yet this itself involves another danger—that of a concentration of material and of casting cavities in the transition areas (Fig. 219*a*). It has, however, been found, from intensive study of the pattern of strength flow in such transition areas, that the lines of force follow a hyperbolic rather than a trigonometrical law. From this we can understand why, in Fig. 92 which shows bone from a human hip-joint, there are no circular arcs, only hyperbolic-type curves, at points of transition. By using curves of this kind in cast-iron constructions, we can substantially reduce the concentrations of material without there being any dangerous increases in stress as a result (Fig. 219*b*). The two parts seen in Figs. 220 and 221—a pressure vessel lid and a guide fork—have been designed along these lines and both are mechanically balanced and extremely sturdy designs.

In the case of machined recessed corners, the radii of curvature will be rather smaller (*b* in Fig. 219). This is connected with the machining process, for large curves are usually produced by means of a template (sometimes even with a *circular turning attachment*), while small ones are always done by eye, which is of course much quicker and cheaper. For this reason small radii are preferred for recessed corners and edges, and the greater notch effect involved is accepted.

With projecting corners and edges things are quite different. In welded constructions these usually remain as they were in the raw material stage (sheet metal or angle-iron). As a rule they are fairly sharp, and if there is a risk of accidents occurring they are rounded off with a file. In castings, edges which

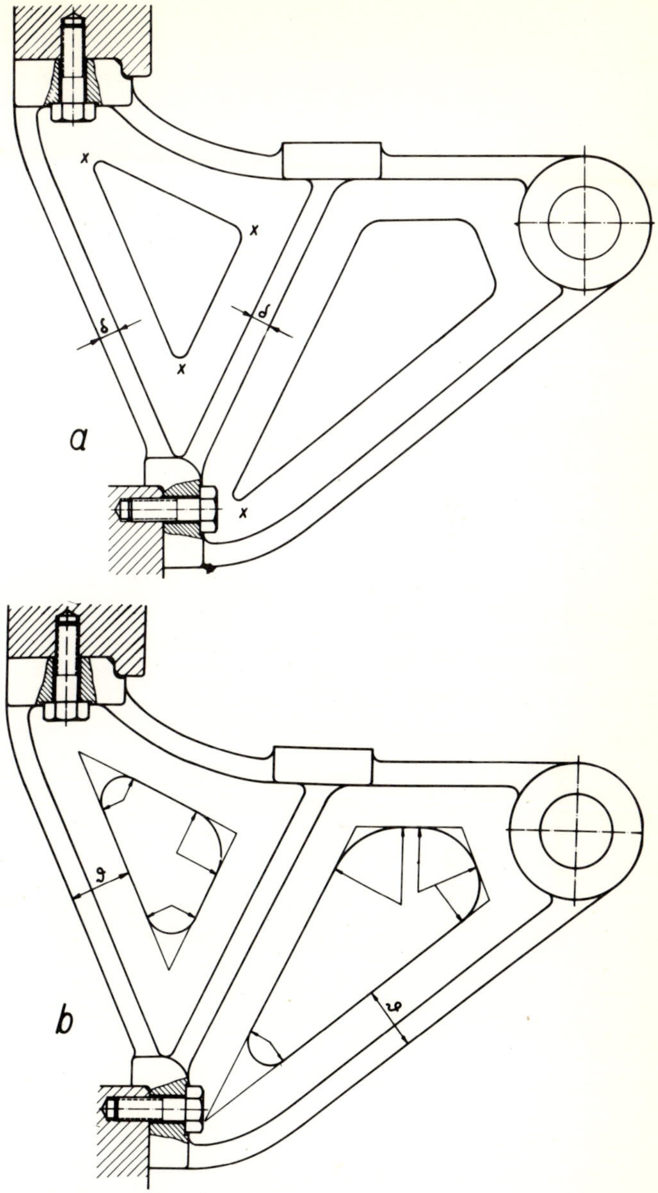

Fig. 222.—Cast bracket with lightening holes. Radii *a* are too small, because they have been calculated on the basis of δ; *b* are correct, being based on ϑ.

project are slightly rounded (*c* in Fig. 212). So far as the lines of force are concerned, their shape makes no difference, since the stresses always run parallel to these edges. Quite apart from this, it would not even be possible to cast sharp edges—at least not in sand—for the probability of the molten metal

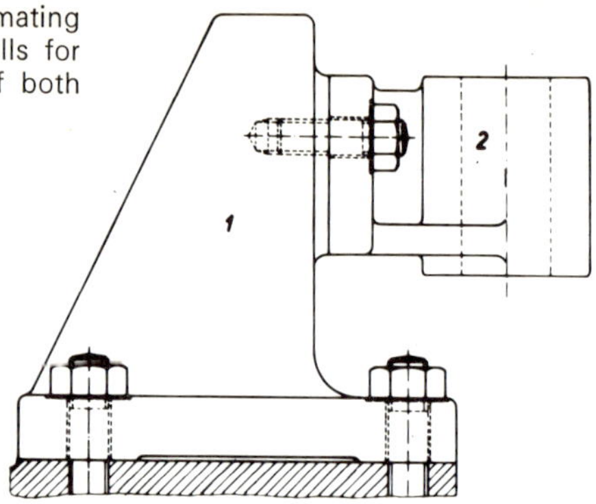

Fig. 223.—The mating of two parts calls for the matching of both surfaces.

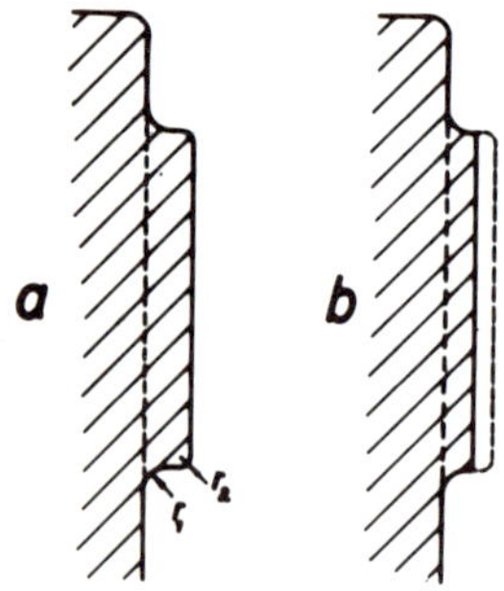

Fig. 224.—(*a*) Initial shape of facing on casting.
(*b*) After machining, with sharp edges.

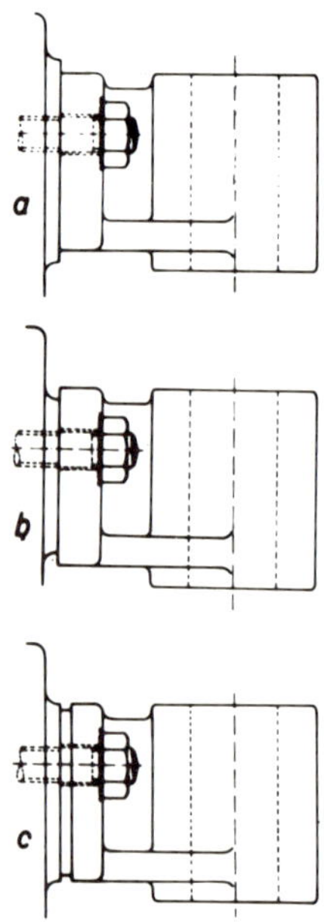

Fig. 225.—Ways of matching a machined facing: (*a*) standing proud, (*b*) set back, (*c*) Brown Boveri method.

hardening prematurely and the mould not being completely filled is too high. This is why it is usual to make projecting edges in castings with a small radius.

On the cylinder cover seen in Fig. 212 we do, however, find one projecting edge which, going against this rule, has a large—indeed, a very large—radius. This is at *g*, where the wall encloses a hollow cavity and the two sides of the corner are at right angles. This is not a projecting edge in the normal sense of the word, but part of a casing which obeys a different rule—that of constant wall thickness, which says that the outer radius R must be larger by the extent of the wall thickness than the inner radius *r* if there is to be no alteration in wall thickness across the whole corner.

Circumstances are totally different where machined projecting edges are concerned. The machining almost always produces a sharp edge (like *d* in Fig. 212). If, when a component is assembled, this edge is covered by a mating part it is left sharp; but if it sticks out—i.e. is unprotected—it will be cut back at 45° to prevent damage occurring. Where a machined projecting edge meets a machined recessed (and therefore radiused) one, the former will be cut back at 45° far enough for there to be no possibility of contact between the two in the radiused area; this would not be suitable for locating the parts in question (*f* in Fig. 212).

Lightening holes in castings represent a rather special type of recessed corner (Fig. 222). If we were to give these a radius calculated by the rules given above (i.e. $r_0/\delta = 0{\cdot}5$ in the case of a right angle), the result would be the sharp-looking corners seen in Fig. 222*a*. In fact a casting like this would not be likely to withstand any severe loads—it would be far too prone to give way at the places marked with an x. Once again, our eye tells us that we would be making a mistake by following an otherwise well-proven rule. It is obvious that here the width governing the stress flow is not just the thickness of the rib δ, but the entire width ϑ; if we replace δ by ϑ in the formulae, they will remain valid. And if we substitute R for r, we can write the equations:

$$\frac{R_1}{\vartheta} = \frac{0{\cdot}5}{1+\cos\gamma} \quad \text{and} \quad \frac{R_2}{\vartheta} = \frac{0{\cdot}5}{1+\cos(180^\circ-\gamma)}$$

The radii shown in Fig. 222*b* have been computed using these formulae, and it will be seen that these 'look right'.

We should mention here one further feature which is found on machine parts of all kinds—the guide facing, or perhaps more correctly, machining facing. Wherever two parts are required to mate in the way shown in Fig. 223, it is essential, if there is to be an exact fit, for the surfaces of both parts to be machined flat. This means that an extra thickness of metal has to be provided on both parts to allow for this machining; in the case of part 2 in Fig. 223 this is done by merely making the foot of the casting rather thicker over its whole width, while with part 1 more metal is added only in the area that will be covered by part 2. This extra thickness is known as a guide facing, or machining facing, and may be round, square, or rectangular. Its height—perhaps it might be better to say its depth—must be sufficient for the machining tool to have an unimpeded run-on and run-off, even if the facing in its rough-cast form does not exactly match its nominal dimensions (this is almost unavoidable in castings and forgings, and within narrower limits in welded assemblies as well). Facings should therefore (depending on the size of the workpiece and the degree of finish required) have a depth of 2–5 mm (0·1–0·2 in.), and in the case of very large work 10–20 mm (0·5–1·0 in.).

Fig. 226.—Discrepancies between the edges of facings are often to be seen in assembled machinery (arrow *a*). A recessed flange can make these less obtrusive (arrow *b*). Turbo-compressor by Brown Boveri, Baden.

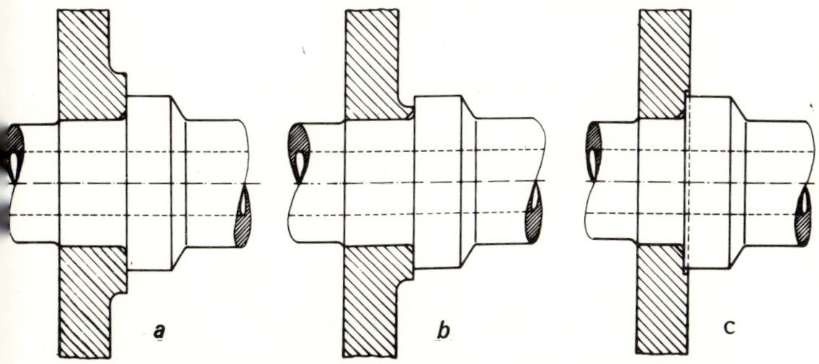

Fig. 227.—When the bearing surfaces are narrow the facing must be made wider (*a*). Otherwise there is a risk of inadequate seating (*b*). An excellent method is to mill the bearing surface into one face (*c*).

Facings will form an integral part of a casting, while in a welded construction they will take the form of plates or rings, directly welded on. Of course, the rules discussed above for edges and corners apply just as much to facings, so the recessed corner at the foot of the facing is given a large radius and the top projecting edges are slightly rounded before they are machined and then left sharp afterwards (Fig. 224); they can be left sharp because they are protected when the parts are assembled. Occasionally, if there is any chance of accidents occurring, they may be cut back with a file.

The width and length of the facing are matched to the dimensions of the mating part. There is no point in making the facing wider or longer, nor is it a good idea to make the facing smaller—in both cases the assembled joint would look as if the parts did not really belong together (Figs. 225 *a* and *b*). But as both shapes never coincide exactly, particularly if they are castings, there will always be slight discrepancies between the facing and the mating part (arrowed *a* in Fig. 226). Although these look better than if there had been no attempt at matching the parts to start with, if the discrepancies are so large as to be obtrusive, they can be removed by grinding, filing, etc. Brown Boveri have found a neat way of avoiding the need for this by recessing the foot of the part offered up to the machined facing, as illustrated in Fig. 225*c*. Discrepancies between the two parts are thus rendered less noticeable (arrowed *b* in Fig. 226).

Exceptions to this rule—identical shapes on both sides of the joint—are made only where the parts make contact over only a very narrow rim to start with (as in Fig. 227*a*); if the machined facing were matched in area to the part mating with it, any eccentricity—which is always a possibility—would result in inadequate contact. The joint might well be functionally unsatisfactory, and would at the very least be unsightly (Fig. 227*b*).

Facings are also provided where, in a completely machined assembly, one type of machining meets another, e.g. where a polished surface makes contact with a ground one. Since the initial machining operation provides good dimensional accuracy, the facing depth can be very shallow, e.g. 0·5 mm (0·02 in.) for normal work and 2–3 mm (0·08–0·12 in.) in the case of large workpieces; an example of this is shown in Fig. 228. These specially machined surfaces include a thread, and it is therefore also usual to machine this onto a facing except where it is milled or cut with a threading die.

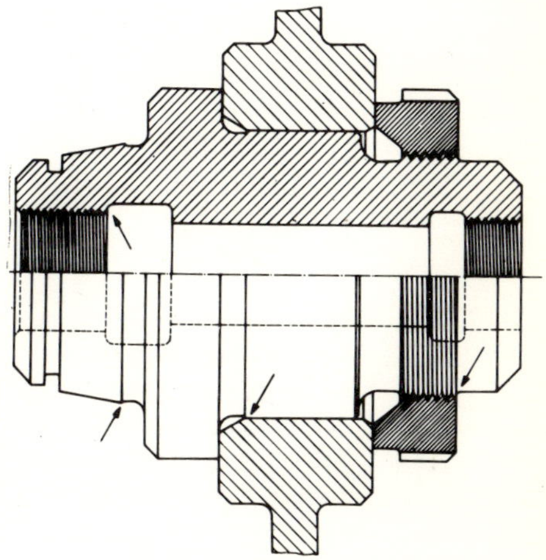

Fig. 228.—Wholly machined workpieces can be made with great accuracy so that quite shallow facings (arrowed) are sufficient.

Situations sometimes arise where there is nothing to be gained from providing a machining facing, and where the run-off for the cutting tool we have mentioned above is not needed; this would, for instance, apply to cases where a centring lip has to be turned into a cylinder wall (*a* in Fig. 229) or axial locating surfaces machined in a circular cavity (*a* in Fig. 230).

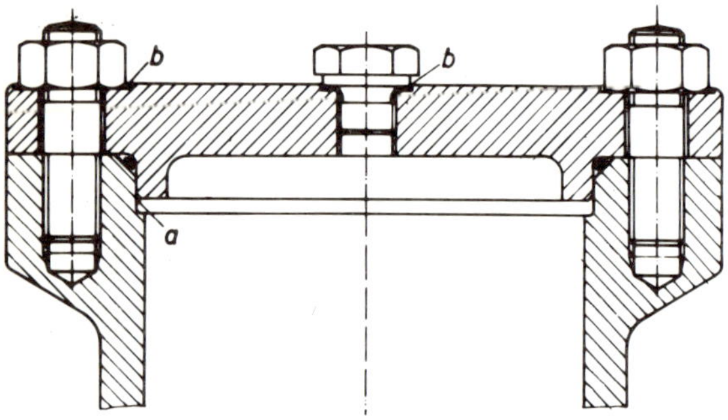

Fig. 229.—Locking cover, with centring lip turned into the cylinder wall. Bearing surfaces are also milled into the top surface for the bolt head and nuts.

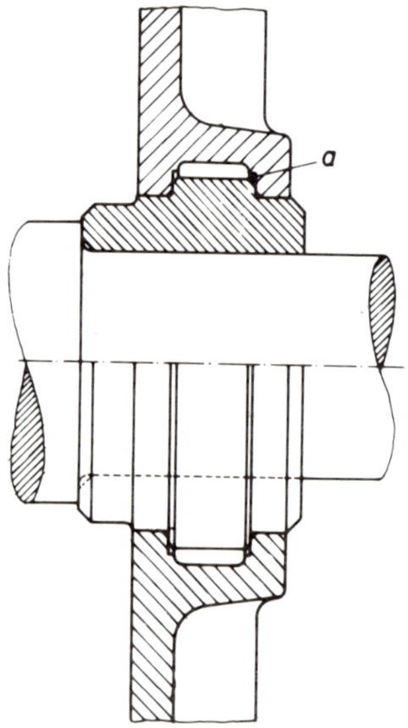

Fig. 230.—Shaft bearing in shell-back form, with axial locating groove in raw casting, without a machining facing.

Similarly, contact surfaces for bolt heads and nuts are not usually machined onto facings but milled direct onto the rough surface of the casting (*b* in Fig. 229). A similar approach is also advisable in cases like Fig. 227, when a special facing is not wanted—as shown in solution *c*.

Designing to match fluid flow

There are so many items of machinery involving a through-flow of gas or liquids that it seems useful to discuss form-design from the point of view of fluid flow theory as well. Pipes and hydraulic units regularly occur in machinery in connection with lubrication, and are often handled by designers who have no specialist knowledge of flow problems; so a general guide may well be found useful.

A fluid flow may take a number of forms, and the first distinction to be made concerns the paths taken by individual particles of the fluid. If these paths follow a regular course, one alongside another, we speak of a *laminar flow*, while if they are haphazard and cross one another, we have what is termed a ***turbulent flow***. Slow-moving flows along narrow passages tend towards the laminar pattern, while rapid flows along a wide channel are more often turbulent. There are, however, other factors which decide whether the flow is laminar or turbulent—in particular, the density d and viscosity η of the fluid. We do not consider them individually, but take a dimensionless quotient known as the *Reynolds number* (Re), arrived at from their various values. The formula for this is:

$$(Re) = \frac{d v l}{\eta}$$

where v=velocity of the flow and l= a reference length (such as diameter in the case of flow in pipes). The flow is more laminar at a low Reynolds number, and turbulent at a high number. In a circular tube the changeover from one type of flow to the other takes place at around (Re)=1700–2000. There are also substantial differences with regard to flow rate distribution (Fig. 231).

With a laminar flow the rate of flow is distributed across the cross-section of the tube according to a parabolic or parabola-like function (I). Where there is turbulence (II) we can distinguish clearly between a central core with a practically uniform and constant flow rate (III*a*), and a kind of envelope in which there is a sharp drop in flow rate towards the wall of the tube (III*b*). This envelope is known as the *boundary layer*,

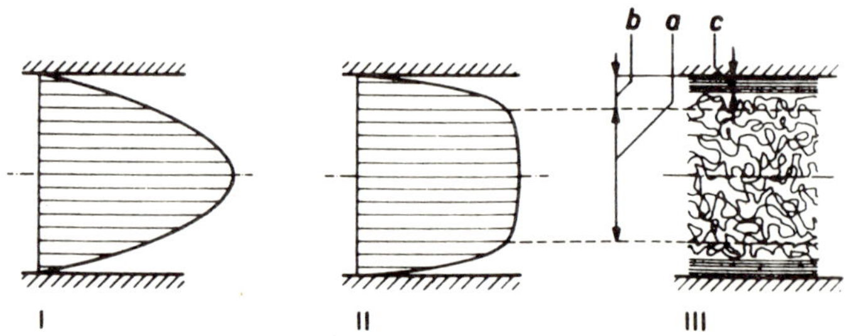

Fig. 231.—Flow rate distribution in tube: I with laminar flow, II with turbulent flow.

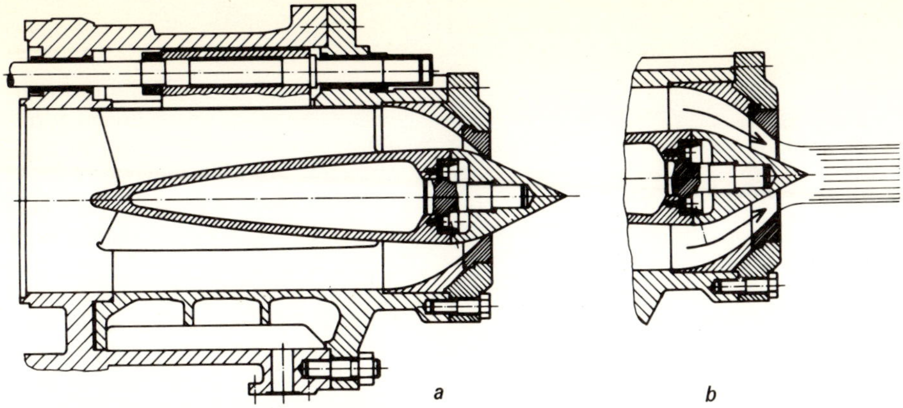

Fig. 232.—Nozzle of a Pelton turbine with needle having longitudinal adjustment in order to vary cross-section of outlet, shown closed (*a*) and open (*b*). The emergent jet reaches its minimum cross-section after leaving the nozzle.

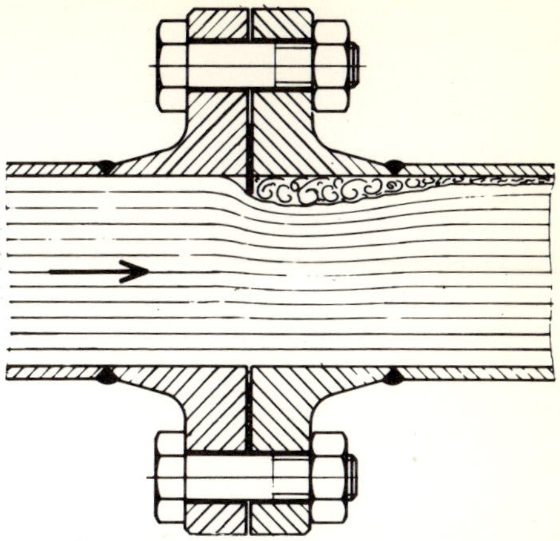

Fig. 233.—Careless assembly of joints in piping may allow the washer to project into the tube and cause boundary layer break-away, with consequent eddy formation and losses.

and is the place where the resistance to flow through friction actually occurs. The outermost regions of the flow boundary layer, next to the wall surface, are always laminar (III*c*); surface irregularities which lie within the laminar boundary layer have virtually no effect on the frictional resistance, since the flow as it were simply passes round them. The faster the flow, the thinner this laminar part of the boundary layer becomes, and the more irregularities protrude into the turbulent part of the flow where they cause friction. Consequently laminar flows are very little affected by roughness of the tube walls, and there is little point in troubling about smooth surface finish; it would be useless to polish the internal surfaces of the spiral casing of a water turbine, for example. On the other hand, there would be a purpose to polishing the sides of the nozzle of a Pelton turbine (Fig. 232) or the vane of a steam turbine. This also means that parts like these should be protected from surface erosion and corrosion, and the rapid increase in energy losses at such surfaces forces the designer more and more to make use of rustproof and wear-resistant materials.

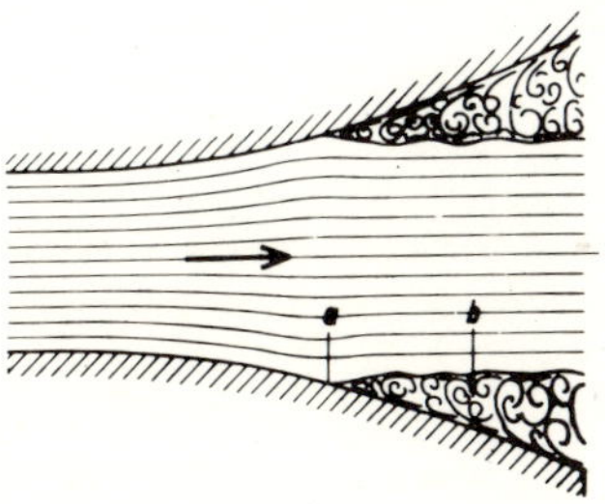

Fig. 234.—Diffuser with increasing divergence caused boundary layer break-away at a certain point, with an associated dead-water zone.

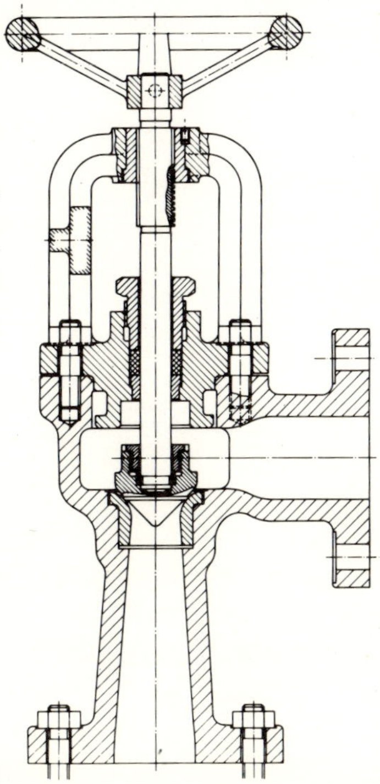

Fig. 235.—Angle valve designed for low flow resistance levels. To avoid break-away in the vertical diffuser, this has been given a small angle of divergence (about 10°).

A particular characteristic of the boundary layer is its vulnerability. Unevenness in the edges of the flow pattern can result in the boundary layer becoming detached from the wall and carried along with the flow. A back-eddy forms behind the point concerned, which may involve a large proportion of the flow and will in any case lead to increased flow losses (Fig. 233). The risk of this boundary break-away is especially high wherever a flow is slowed down, and where as a result the pressure rises in the direction of flow (Bernoulli's principle). The thrust exerted by the boundary layer is not enough to overcome a substantial increase in pressure, and as a result the boundary layer comes to a halt, and a back-eddy occurs which penetrates beneath the boundary layer from the opposite direction and lifts it off the wall surface (Fig. 234). The associated eddy zone produces an intense exchange of momen-

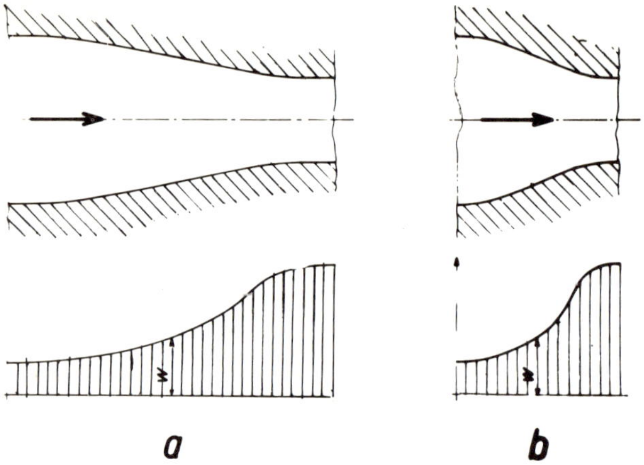

Fig. 236.—Long and short ram tubes. There is no risk of boundary layer break-away, so there is no need for a long run-in. Extensive contact areas between wall and flow cause high friction losses at the higher flow rates (*a*), so that a short length (*b*) is better.

Fig. 237.—The higher the rate of flow, the larger the radii of the walls need to be. For this reason the best shape for the lip of a tube is not an arc (*a*), but a spiral as shown in (*b*). If boundary break-away occurs, it will be much later with (*b*) than with (*a*).

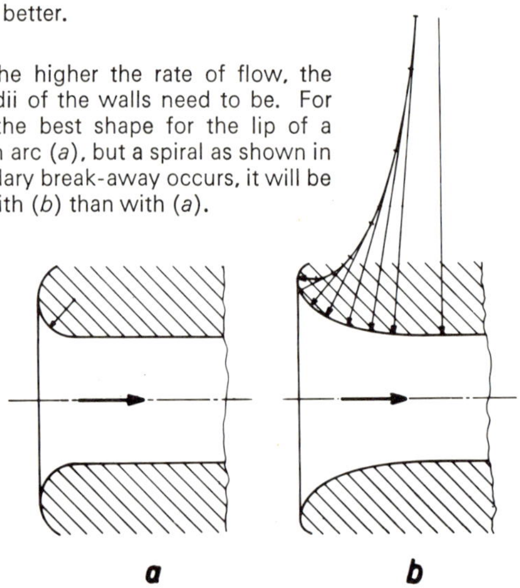

tum with the main flow which, as we have said, loses a large part of its energy as a result. It is very difficult to prevent this break-away from happening where a tube or passageway widens out—in fact, it is only possible when the degree of divergence is very slight. An angle of 10° can be taken as a reasonable average (Fig. 235).

On the other hand it is quite a simple matter to speed up a fluid flow. Although the decrease in cross-section should be an even one, it can take place over quite a short length (Fig. 236). Short acceleration distances are to be preferred to long ones, because when the flow reaches high speeds it is subject to substantial friction losses through contact with the wall; this explains the very stubby design of the Pelton nozzle seen in Fig. 232.

There is a particular tendency for boundary layer break-away to occur along curved sections of wall, where the flow is driven outwards, away from the wall, by inertia. This should be remembered when shaping the mouth of a tube or passageway. The higher the flow rate, the larger the radius of curvature of the walls should be; and the best shape for the mouth of an intake is not an arc of a circle, but a curve with a constantly increasing radius (as illustrated in Fig. 237). There is a parallel here with the pattern of stress flow at areas of transition (cf. section on 'Designing to match the stress flow').

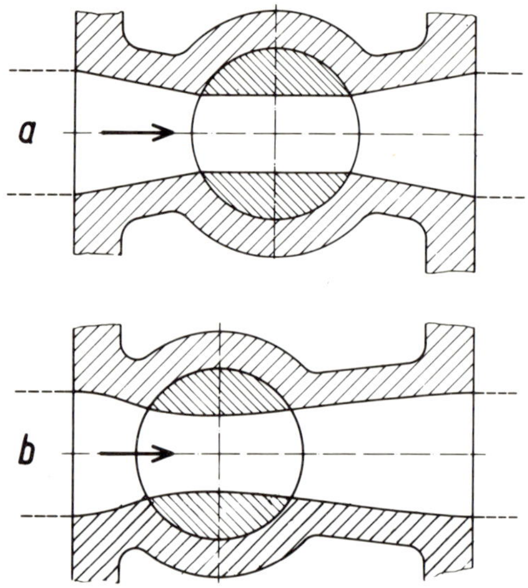

Fig. 238.—(*a*) Wrong and (*b*) right shape for a cut-off valve. Instead of having an equal length for the inlet and outlet, the narrowest cross-section should be moved upstream. The transitions should also be smoothed off.

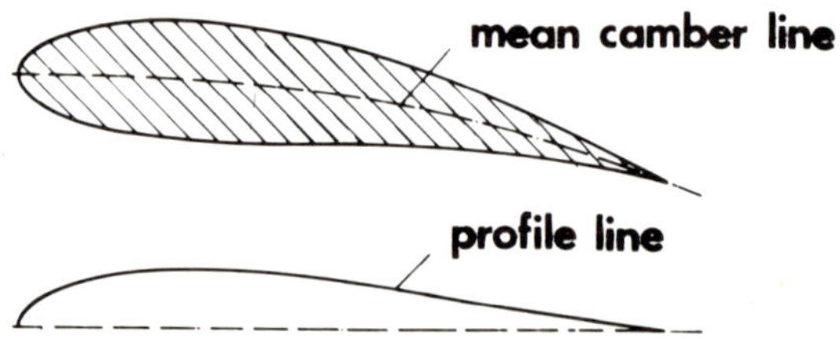

Fig. 239.—An aerofoil shape can be broken down into two components, known as the mean camber line and the profile. In seeking the ideal shape for the flow involved, these components should be studied separately.

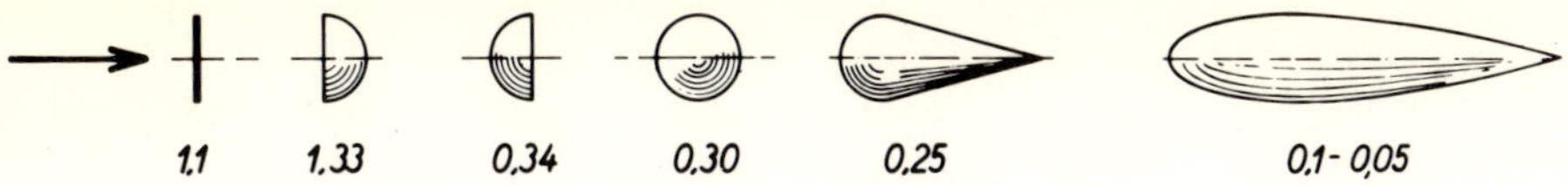

Fig. 240.—Bodies of revolution with various flow resistance values. The figures indicate the coefficient of resistance.

From all this it will be obvious that any bend in a tube will bring about extra movement of the fluid passing through it. If the bend is too sharp, there is a danger of break-away along the inner edge; if it is too shallow, the wall friction will be too high. And so a compromise has to be found between the two extremes. One way or the other, a loss in pressure is inevitable; in a first approximation, this loss can be taken as proportional to the angle of deflection.

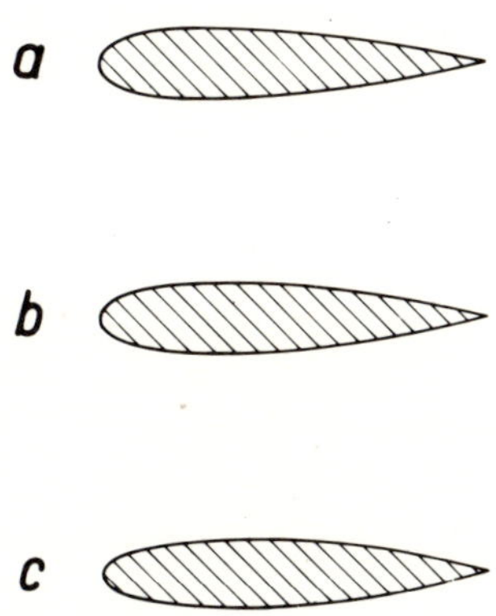

Fig. 241.—The best profile varies with the kind of flow. (*a*) is suitable for low Reynolds numbers, (*b*) for medium and (*c*) for high numbers.

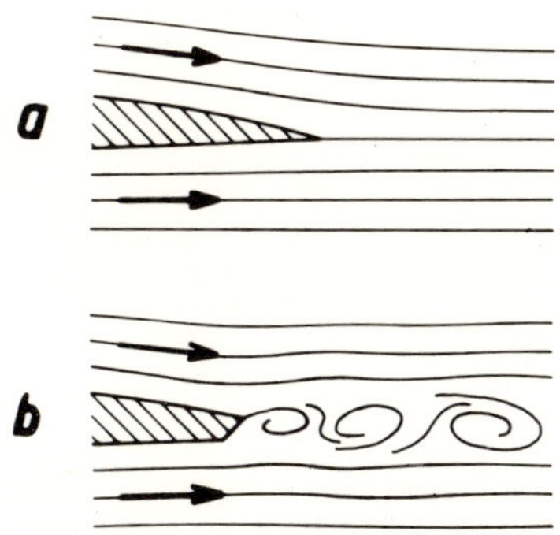

Fig. 242.—The trailing edge of a vane or wing should allow the two parts of the flow to meet smoothly (shape *a*). In practice, however, a different shape is often unavoidable (*b*), and this causes turbulence and losses.

The swirl which a bend imparts to a fluid flow can often still be detected several diameters further on along the tube, and for this reason accelerating nozzles should never be situated immediately after a bend, but be preceded by a certain straight length to allow the flow to settle down. The length needed for this can be reduced by fitting a smoothing grid in the path of the fluid; a grid of this kind can, for instance, be seen in the arms of the cross supporting the nozzle needle of the Pelton turbine shown in Fig. 232.

To sum up, we can say that every change in shape, and every deflection will adversely affect a fluid flow. Projecting edges and facings on the walls along which the flow passes should be rounded (or, indeed, completely avoided if at all possible). They cause a good deal more trouble than sharp recessed corners. This agrees entirely with what we have already said about the rules laid down on stress flow, except that there it was the recessed edges that had to be radiused and not the projecting ones. In both cases, the edges cause disruption of a flow pattern; and so when we have a part that forms a boundary for high stress on the inside and carries a fluid flow on the outside, it must have a smooth and flowing shape both internally and externally. The shut-off valve seen in Fig. 238, for example, should be shaped as shown in *b*, and not as in *a*.

The same considerations apply to parts which are placed in a flow in order to draw energy from it or impart energy to it, e.g. the vanes of pumps, turbines and compressors. The way these work, and what their energy losses are, also depend to a very great extent on their external shape. This can in all cases be broken down into two components, known as the *mean camber line* and the *profile* (Fig. 239). To enable a vane to transfer energy, it has a curved mean camber line; and as a rule we try to distribute the curvature evenly over its length (i.e. make it an arc of a circle). Finding a suitable profile is very largely an empirical matter; it is quite amazing how much the resistance

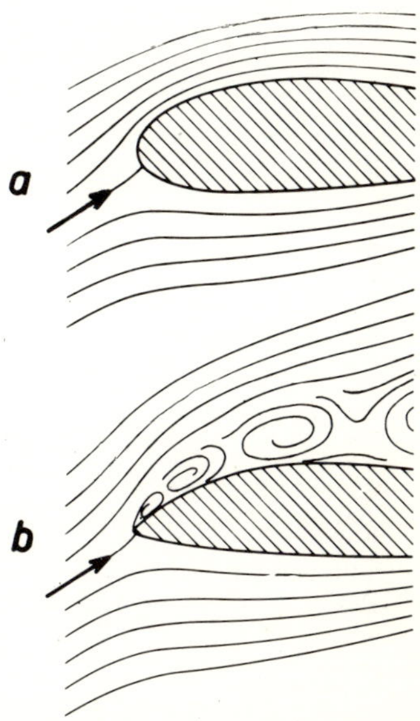

Fig. 243.—A rounded leading edge (*a*) is less sensitive to a wrong angle of incidence than a pointed one (*b*).

to flow of different shapes can be, and Fig. 240 compares a number of shapes and shows their coefficients of resistance. It is important to note that the ideal profile may vary for different kinds of flow, the best shape varying with the Reynolds number. The higher this is, the more pointed the leading edge of the profile needs to be and the further back towards the centre of the vane the thickest point must be shifted (Fig. 241).

We try to achieve a profile which will avoid any boundary layer break-away, or at least move this as far back as possible. The trailing edge should be as sharp as possible, so that the two parts of the flow (upper and lower) come together smoothly (Fig. 242*a*). For production reasons, however, we are often obliged to cut off the rear edge of the profile more or less abruptly, which leads to a break-up of the smooth flow and to eddy losses (Fig. 242*b*).

People are often surprised, however, that the leading edge may on the contrary be blunt or rounded, and they feel that a sharp edge would be more correct. The very rounded nose given to an aircraft fuselage, for example, does indeed appear to be a simplification of the flow requirements. And yet there are actually good reasons for preferring a blunt point to a sharp one—the flow is slowed down over the nose and an almost stationary wedge of air builds up which splits up the air passing over it. Moreover, when the flow does not meet the leading edge exactly at a tangent to the mean camber line, there is less boundary break-away over a rounded edge than there would be with a sharp one (Fig. 243).

index

aerofoil, 110-11
aesthetics, 16-19

beam,
curved, 81-2
sections, 73
bearing,
friction, 12
pedestal, 25
thrust, 11
beauty, 17-18
belt drive, 5
bending
resistance, 72-3
stresses, 89
Bernoulli, 109
brazing, 41
British Standards, 24
limits and fits, 37
roughness comparison, 31
surface texture, 29
buckling, 70-1
of webs, 78
calculation and design, 9
camber line, 110
Carnot cycle, 9
casting, 46
chamfer, 93
chassis, lorry, 100
circular tube,
bending, 71
flow in, 108
torsion, 73-4
clearance fit, 38
cold-drawing, 46

compression sections, 72-3
concentration, stress, 53, 88
cone locking ring, 7
constant,
  proportion law, 25
  wall thickness, 61-70
control, 53
corner shapes, 102-6
courage, 14-16
crane design, 73
curved beams, 81
cut-off valve, 110

density, fluid, 108
design,
  and aesthetics, 16-19
  and calculation, 9
  and drawing, 19-22
  details, 12
  early, 16-17, 20-4
  form, 10, 57
  moderation, 10
  nature of, 1-2
  patent, 6
  specification, 7
  stage, 3
  steam rotors, 2
  value of, 1-2
details, design of, 12
Diesel engine,
  crank, 13
  design, 84
diffuser, 109
drawing, 19-22
  workshop, 21
drive, belt, 5-6
drop-forging, 46

early design, 16-17, 20-4
eddy, 109
Edison, Thomas, 4
elements,
  of design, 57-61
  machine, 3, 7
engine crank, 13
extrusion, 48

facings, design of, 107
failure, 90, 96
fatigue, 90, 94
finish, surface, 29
fits, 37-8
  clearance, 38
  interference, 38
  running, 38
  shrink, 13, 98
  tight, 38
  transition, 38
flange, 63
flow,
  fluid, 108-112
  laminar, 108
  strength, 83, 92
  turbulent, 108
fluid flow, 108-112
fracture, 90, 96
  surfaces, 91
friction bearing, 12
frictional resistance, 109
fundamental tolerance, 37

gauges, 36-7
gland design, 13
go and no go, 36
Grob process, 49

hand-working, 42-5
hard-soldering, 41
hole, lightening, 63, 78, 105
Hyde's hard-soldering, 41

inertia, moment of, 71
injection moulding, 51
interchangeability, 35
interface notch, 97
interference,
  fit, 38
  microscope, 32
International Standards, 24
invention, 3-6

keyway, 94
knee-plate, 82, 98
Krupp, 20

laminar flow, 108
Leonardo da Vinci, 20
lightening hole, 63, 78, 105
limits, 36
lines of force, 82-3
locking ring, cone, 7
lorry chassis, 100
lubrication, 11

machine elements, 3, 7
machining, 46
mean tolerance, 38
moderation, art of, 10
moment
  of area, second, 71
  of inertia, 71

no go, 36
nominal size, 35-7
notch,
  interface, 97
  relief, 93
  sensitivity, 89
  stress, 87-9
  stress-relieving, 92
numbers,
  preferred, 24
  Renard, 26
  Reynolds, 108

patent, 4
  design, 6
  lapse of, 5
pedestal bearing, 25
Pelton turbine nozzle, 109
perseverance, 8
Perth-O-Meter, 31
plate,
  in torsion, 75
  knee-, 83, 98
plug gauge, 36-7
Prandtl, 94
preferred numbers, 24
production processes, 40-52
profile, fluid flow, 110-111
proportion, law of constant, 25
pump, bore-hole, 7

quality, surface, 37

radius, transition, 104-5
railway locomotive, 16-17

ram tube, 110
relief notch, 93
Renard numbers, 26
resistance,
  to bending, 71
  to buckling, 71-3
  to torsion, 71-4
Reynolds number, 108
ribs,
  in torsion, 75
  stiffening, 64-7, 71
Riedler, 19, 20
ring, cone locking, 7
riveter stand, 86-7
rotors, 1
roughness, surface, 29
running fit, 38

screw threads, 93
second moment of area, 71
sections,
  for compression, 72-3
  for torsion, 72-3
self-tensioning drive, 6
sensitivity, notch-, 89
shapes,
  effect on loading, 83-9
  strong and weak, 70
shear stress, 89
shearing of webs, 78
shrink fit, 13, 98
snap gauge, 36-7
soap bubble analogy, 94
specification,
  design, 7
  standard, 7
spline shaft, 95
standard specifications, 7
  British, 24
  international, 24
standardization, 22-6
steam rotors, 2
stiffening ribs, 64-7
  in torsion, 77
Stodola, 2
strength,
  bending, 89
  flow, 82, 92
  uniform, 10
stress concentration, 53, 88
  index, 89
  notch theory, 87-9
  relief notch, 92
  shape effect, 83-9
structure, surface, 27
surface,
  effect on strength, 92
  finish, 29
  form, 27
  position, 34
  quality, 37
  roughness, 29
  shaping, 106
  structure, 27
  texture, 27
  waviness, 29
swaging, 78

Tacoma bridge, 77
Talysurf, 31
technical feel, 10
testing, 53-4
texture surface, 27
  British Standard, 29
theory,
  notch stress, 87-9
  of surfaces, 26-40
Thum, Professor, 57, 89
thrust bearing, 11
tight fit, 38
tolerance, 35-7
  fundamental, 37
  mean, 38
  unit, 37
torsion, 71-81, 99
  plate in, 75
  sections, 72-3
tractor chassis, 101
transition,
  fit, 38
  radius, 104-5
transport problems, 2-3, 15
tubes, 71-2
  ram, 110
tubular construction, 71-2
turbine,
  nozzle, 109
  rotor, 2
twisting, 71-81, 99
uniform strength, 10
unit-
  bore, 40
  shaft, 40
  tolerance, 37

valve, cut-off, 110
Vinci, Leonardo da, 20
viscosity, fluid, 108

wall thickness, constant, 61-70
waviness, surface, 29
web,
  buckling, 78
  shearing, 78
welding, 97, 102-3

yielding, 89

Zeiss interference microscope, 32